Global Genes, Local Concerns

Global Genes, Local Concerns

Legal, Ethical, and Scientific Challenges in International Biobanking

Edited by

Timo Minssen

Centre for Advanced Studies in Biomedical Innovation Law, University of Copenhagen, Denmark

Janne Rothmar Herrman

Centre for Advanced Studies in Biomedical Innovation Law, University of Copenhagen, Denmark

Jens Schovsbo

Centre for Information and Innovation Law, University of Copenhagen, Denmark

Cheltenham, UK • Northampton, MA, USA

Published by
Edward Elgar Publishing Limited
The Lypiatts
15 Lansdown Road
Cheltenham
Glos GL50 2JA
UK

Edward Elgar Publishing, Inc.
William Pratt House
9 Dewey Court
Northampton
Massachusetts 01060
USA

A catalogue record for this book
is available from the British Library

Library of Congress Control Number: 2018967809

This book is available electronically in the **Elgar**online
Law subject collection
DOI 10.4337/9781788116190

ISBN 978 1 78811 618 3 (cased)
ISBN 978 1 78811 619 0 (eBook)

Printed and bound in Great Britain by TJ International Ltd, Padstow

Contents

Contributors

Human Biosample Governance Professional Tina Bossow, Novo Nordisk A/S

Senior Scientific Director Brian J. Clark, Global Head of Biobanking, Novo Nordisk A/S

Professor Timothy Caulfield, Health Law Institute, University of Alberta

Associate Professor Åsa Hellstadius, Department of Law, University of Stockholm

Professor Janne Rothmar Herrmann, Centre for Advanced Studies in Biomedical Innovation Law (CeBIL), University of Copenhagen

Professor Klaus Hoeyer, Section of Health Services Research, University of Copenhagen

Research Assistant Matthew Jordan, Centre for Law, Medicine and Life Sciences, University of Cambridge

Professor Jane Kaye, Centre for Health Law and Emerging Technologies, NDPH, University of Oxford and Melbourne Law School, University of Melbourne

Postdoctoral Researcher Nana Cecilie Halmsted Kongsholm, Department of Media, Cognition and Communication (MCC), University of Copenhagen

Senior Lecturer Kathleen Liddell, Director Centre for Law, Medicine and Life Sciences, University of Cambridge

Senior Research Associate Johnathon Liddicoat, Centre for Law, Medicine and Life Sciences, University of Cambridge

Professor Michael J. Madison, Institute for Cyber Law, Policy, and Security, University of Pittsburgh

Professor Timo Minssen, Managing Director, Centre for Advanced Studies in Biomedical Innovation Law (CeBIL), University of Copenhagen

Research Associate Blake Murdoch, Health Law Institute, University of Alberta

Doctor Eva Ortega-Paíno, Coordinator for the BBMRI.se Service Centre for Southern Sweden, Lund University

Assistant Professor William Nicholson Price II, University of Michigan

Research Associate Megan Prictor, Melbourne Law School, University of Melbourne

Research Associate Malene Bøgehus Rasmussen, Section of Health Services Research, University of Copenhagen

Professor Karine Sargsyan, Medical University of Graz

Professor Jens Schovsbo, Centre for Information and Innovation Law (CIIR), University of Copenhagen.

Associate Professor Aaro Tupasela, Section of Health Services Research, University of Copenhagen

Professor Franziska Vogl, Medical University of Graz

Assistant Professor Helen Yu, Centre for Advanced Studies in Biomedical Innovation Law (CeBIL), University of Copenhagen

Professor Peter K. Yu, Texas A&M University

Professor Esther van Zimmeren, Research Group Government and Law, Faculty of Law, University of Antwerp

Introduction

MUCH ADO ABOUT SOMETHING: LEGAL, ETHICAL, AND SCIENTIFIC CHALLENGES IN INTERNATIONAL BIOBANKING AND TRANSLATIONAL EXPLOITATION

The significance of large-scale, interoperable biobanks as knowledge institutions and research infrastructures in today's life science research can hardly be overestimated. Biobanks allow researchers with different scientific expertise to analyse large and diverse collections of human biological material (HBM), as well as genetic, clinical, health, and other personal data of donors.

Hence, it is no surprise that multiple types of biobanks are being established around the globe with very different financial, organizational, and legal setups. These include biobanks of various sizes that may be disease-centric, population-based, genetic or DNA/RNA, project-driven, tissue type, multiple specimen type, commercial, or virtual biobanks.[1] While their aims and ambitions may vary considerably, many are created to operate for several decennia and with the aim to provide a valuable resource and infrastructure that can be accessed by numerous research projects and by a great variety of stakeholders with different objectives. These may include private companies, university researchers, research foundations, governmental bodies, or "hybrid" consortia in the framework of private–public partnerships (PPPs).

The increasing significance and complexity of biobanking requires substantial investments in the creation, organization, and maintenance of collections of HBM and the "big data" stored in biobanks. This highlights the importance of effective governance and use of biobanks and raises the question of how to deal with a great variety of scientific, ethical, and legal challenges.

This book encompasses a broad range of chapters written by experts that address and discuss some of these challenges from an interdisciplinary perspective. These chapters have been prepared as part of the research project *Global*

[1] De Souza, Y. G., & Greenspan, J. S. (2013). Biobanking Past, Present and Future: Responsibilities and Benefits. AIDS (London, England), 27(3), 303–312. http://doi.org/10.1097/QAD.0b013e32835c1244.

Genes, Local Concerns.[2] Reflecting the goals of the crossfaculty research project, the project dealt with the legal, ethical, and scientific challenges in cross-national biobanking and translational exploitation. In the present book, leading international researchers discuss pressing questions, such as: How do national biobanks best contribute to translational research? What are the opportunities and challenges that current regulations present for translational use of biobanks? How does inter-biobank coordination and collaboration occur on various levels? How could academic and industrial exploitation, ownership and IPR issues be addressed and facilitated?

Special emphasis was placed on legal and ethical challenges and opportunities in addressing regulatory barriers to biobank research and the translation of research results, while at the same time securing the ethical legitimacy of the research and the societal interests in access to information and innovation.

A recurrent theme in the project which is also reflected in the chapters has been the issue of safeguarding autonomous decisionmaking. This is a classical issue in health law and medical ethics that has been revitalized in the setting of international biobanking, where samples are collected and shared across borders and therefore across different ethical and legal regimes. The legal frameworks that apply to biobanks are fragmented, vary from country to country and change over time. They are only in some countries specific to biobanks, and in most countries regulate data protection, privacy or human research generally. Yet autonomous decisionmaking is a cornerstone in most if not all Western legal systems. Equally, consent—be it blanket, broad, or informed—is of central ethical importance in the context of medical research, including biobank research. Accordingly, much attention was given to developing appropriate consent and patient involvement models that adequately protect the interests of biobank donors.

But legal, ethical, and scientific challenges also arise from rapid technological developments, in which area more and more novel uses are emerging. The complexity of these challenges is increased by the growing need to secure funding for research and for the maintenance and collection of samples at high quality facilities. Hence, additional issues have to be dealt with, ranging from obligations regarding how to properly share data and issues of ethics to commercial, legal and trust-related issues in translational medicine, and tech transfer. This also involves the discussion of principles for biobank guidelines, intellectual property policies, (open) collaboration, and governance. To prop-

---

2 For further information see http://globalgenes.ku.dk. The project was funded by the UCPH Excellence Programme for Interdisciplinary Research, Principal Investigator: Timo Minssen; see http://research.ku.dk/strengths/excellence-programmes (both accessed 15 February 2019).

erly address this broad range of issues, the collection of contributions that were complied for this volume have been organized into four parts and 14 chapters.

Part I of this book is devoted to issues revolving around biobanks, Big Data, and modes of collaboration.

In Chapter 1, *Klaus Hoeyer*, *Aaro Tupasela*, and *Malene Bøgehus Rasmussen* point out that cross-national collaboration in medical research has gained increased policy attention. In particular, the authors explain how various policies are developed to enhance data sharing, ensure open access, and harmonize international standards and ethics rules in order to promote access to increase scientific output and facilitate more data-intensive research, including with what is sometimes referred to as Big Data. In tandem with this promotion of data sharing, numerous ethics policies are developed to control data flows and protect privacy and confidentiality. According to the authors, both sets of policymaking however pay limited attention to the moral decisions and social ties enacted in the everyday routines of scientific work. Using the example of practices of a Danish laboratory with great experience in international collaboration regarding genetic research, the authors focus on "ethics work" and argue that it is crucial for data sharing, though it is rarely articulated in ethics policies and remains inadequately funded.

In Chapter 2, *Michael Madison* takes a more general perspective. He considers biobanks as a case of knowledge commons, or collaborative institutions designed for the preservation and generation of knowledge. He explains how analyses of knowledge commons often focus on legal and contractual dimensions of openness and access to shared commons resources. It is important, however, to put those analyses in a broader context. Given the number and diversity of biobank institutions worldwide, and the critical role sometimes played by public policy and state support in ensuring their continued existence and service, it is pointed out in this chapter that it is important for researchers to consider, empirically and in a systematic manner, all relevant aspects of biobank governance, including their shared sources of strength and weakness and particular sources of opportunity and concern.

Part II of this book focuses on biobanks, translational medicine, and tech transfer.

In Chapter 3, *Nicholson Price* explains and discusses the importance and relevance of the Big Data dimension of biobanking for the promotion of translational medicine. The chapter is divided into two parts. The first briefly describes the sources of medical data, the promises of medical big data, and the key challenge of data fragmentation. The second discusses the role of biobanks in medical big data, focusing on their role in infrastructure for innovation and their potential for facilitating translational research. The author argues that viewing biobank-related data as infrastructure would place them at a distinctly earlier point in the commercialization pipeline, serving more to facilitate later steps in translational medicine rather than being viewed as potentially commercializable products themselves.

In Chapter 4, *Brian J. Clark* and *Tina Bossow* explain and discuss the need of the bioscience industries to gain access to biosamples in order to boost their research and development (R&D) capabilities and hence to enhance the chances of translational innovation. But the authors also highlight that it is important that both sourcing and use are responsible and well governed. In their view, a pivotal question for access is whether the bioscience industries can demonstrate that they use human biological samples (HBS) and data in ways that are socially responsible, ethical, compliant with applicable laws or regulations, and safe. Moreover, the authors stress that this same question should equally apply to any user of HBS, whether a for-profit industrial user or a nonprofit or public institutional user.

This leads us to Part III, which addresses central issues relating to the interface of biobanks, human rights, and patient involvement.

In Chapter 5, *Peter Yu* starts by providing a brief survey of three distinct sets of human rights issues that are all related to biobanks. The first set concerns the human rights involved in the collection, processing, use, or storage of the biological materials collected by biobanks. The second set pertains to the human rights issues implicated by the development of scientific productions utilizing the collected materials. The third set relates to the human rights obligations of three types of biobanks: public biobanks, private biobanks, and biobanks formed out of public–private partnerships (PPPs). The author stresses that the goal of this chapter is not to provide detailed analyses of these three sets of human rights issues; instead, the chapter aims to offer preliminary sketches of the various human rights issues that can be implicated by biobanks.

Åsa Hellstadius and *Jens Schovsbo* argue in Chapter 6 that patent law should be understood in light of health law and human rights, thus highlighting free and informed consent as a vital issue that falls under the morality exclusion enshrined in European patent law. The authors argue that consequently compliance with requirements of free and informed consent should be monitored *ex officio* by patent authorities, and could in rare circumstances lead to nonpatentability. In their view, there is no doubt that this avenue, where protection of donor rights is tied directly to the commercial potential, would give added gravitas to free, informed consent.

In Chapter 7, *Jane Kaye* and *Megan Prictor* demonstrate how a technological platform can support individual decisionmaking in making consent dynamic. It allows for the ongoing engagement of donors in a way that reflects their personal preferences. In the context of biobanking, the authors explain how a Dynamic Consent tool may enable potential participants to give broad consent or to specify in advance that their permission must be sought for each new use of their samples or data. They can revisit and change these specifications over time, including withdrawing their consent and being assured that this has taken effect.

In Chapter 8, *Esther Van Zimmeren* explores whether "Dynamic Consent" could be an effective tool to increase transparency and trust in general, and more specifically regarding commercialization issues. She starts with a short description of the trust literature and tailors some important concepts from that literature to the discussion on biobanks before delving in more detail into the literature on biobanks and trust and the potential role of Dynamic Consent for generating trust. The author acknowledges the link between interpersonal and organizational trust, which seems to be quite critical within the context of biobanks. Moreover, she highlights the intricate dynamics and potential roles of particular persons and third parties in influencing the level of organizational trust in biobanks when they use Dynamic Consent interfaces.

Next, *Nana Kongsholm* argues in Chapter 9 that efforts to harmonize informed consent requirements in international biobanking risk overlooking local factors that may compromise free and informed consent, particularly when research is conducted in developing countries. Drawing on findings from an interview study with donors in rural Pakistan, the author demonstrates how psychological, cultural, and structural factors in this particular context may pose serious ethical challenges that are far from adequately accommodated by (and may in fact thrive under) any standard consent scheme. She argues that vulnerabilities to exploitation in research are highly dependent on social conventions. Hence, customs and harmonization efforts should be supplemented with appropriate efforts to highlight and accommodate such vulnerabilities.

Chapter 10, by *Tim Caulfield* and *Blake Murdoch*, also points to the fact that there are numerous social forces and cultural trends that may be intensifying unresolved consent issues, while acknowledging the practical needs that drove the adoption of the modified consent strategies. They argue that researchers, participants, and institutions would all benefit from a defensible, sustainable, and conceptually coherent consent policy. Given the rise in privacy concerns, the increased interest in rights of control, the rapid pace of technological development, and the lack of consensus on preferred consent type, the authors urge policymakers and politicians to clear up the confusion.

Finally, Part IV of the book is devoted to biobanks, guidelines, and good governance.

In Chapter 11, *Helen Yu* points out how one of the core objectives of responsible research and innovation (RRI) is to maximize the value of publicly funded research so that it may be returned to benefit society. In the case of biobanking, however, the personal nature of human biological materials and potential altruistic intentions of participants to donate samples intensifies the need to adhere to RRI principles with respect to the research, development, and commercialization of innovations derived from biobanks. To bridge the seemingly contradictory and competing objectives of open science and commercialization, the author proposes a holistic innovation framework directed at improving RRI practice to obtain the optimal social and economic values from research.

Next, *Franziska Vogl* and *Karine Sargsyan* explain in Chapter 12 how access to long-term funding for biobanks is still an issue and strategies to recover biobanking costs are emerging. The usage of all collected samples, or use of the well-functioning and expensive infrastructure only for one project collection, is unusual. Instead, a considerable number of biobanks are opting for diversity and running additional population-based collections almost without any time limitations for retrospective and epidemiological studies. That means that research usage of biospecimens is unspecified in terms of time and matter. This, however, is often untenable and the authors point out that those involved in biobanks need to ask themselves how much it will cost to terminate themselves, should it become necessary.

In Chapter 13, *Eva Ortega-Paino* and *Aaro Tupasela* refer to the BBMRI-ERIC's experiences and structures to ask in what ways biobank networks can facilitate sharing, not only of samples but also of information for improving and tackling diseases. The authors further stress that biobank networks give rise to new governance structures in which new ethical and legal (soft law) norms are established and exercised. These norms have considerable implications in relation to how we perceive the acceptability of new practices regarding the collection, distribution, and use of biobanking samples. The authors finally point out how biobanks and biobank networks play a crucial role in maintaining the social and technical norms that allow for tissue economies to emerge and function.

Finally, Chapter 14, authored by *Kathleen Liddell*, *Johnathan Liddicoat*, and *Matthew Jordan*, brings "the issue of IP policies for large bioresources out of the long shadows of rhetoric about openness." In doing the so, the authors highlight two fictions: *first*, that the idea of openness is clearly defined; *second,* that organizations are committed to openness. At the same time the chapter emphasizes that the "harmonization of bioresources' access policies" is a feasible and desirable goal. The authors conclude by outlining future research to improve openness and IP policies for large bioresources.

Having reached the end of the *Global Genes, Local Concerns* project, it is our hope that our publications and the chapters we compiled for this volume contribute to the continued development of international biobanking by highlighting and analyzing the complexities in this important area of research. We also hope that this volume and the challenges that it highlights help to raise greater interest and attention from the medical community biobank operators and funders, policymakers, regulators, commentators, and the mass media.

The complexity of the issues touched upon indicates that many questions remain unsolved. Hence, we are very grateful for all the great presentations and inspiring interactive panel debates seen during the project, which provided more than enough fodder for future research projects. The constructive comments and questions we received from the multistakeholder participants in the project also demonstrated how important further research is in this area.

This again reminds us of the fact that biobanks are providing an increasingly important research infrastructure not only for biomedical researchers that are working *with* them, but also for the social, economic, and legal scientists conducting research *on and about* biobanks. It remains a vital task for we scientists to help clear up the sometimes opaque and elusive legal and economic challenges of international biobanking, in order to provide the basis for better access to high-quality research material and better use and sharing of these essential resources.

Copenhagen, July 1, 2018
Timo Minssen/Janne Rothmar Herrmann/Jens Schovsbo

PART I

Biobanks, Big Data, and modes of collaboration

1. Big Data and the ethics of detail: the role of ethics work in the making of a cross-national research infrastructure for genetic research

Klaus Hoeyer, Aaro Tupasela, and Malene Bøgehus Rasmussen[1]

1 INTRODUCTION

This chapter explores the mobilization of resources, technologies, patients, human capital, and biomaterials for international collaboration regarding genetic research. Our question is simple: What makes genetic material and health data flow, and which hopes and concerns travel along with them? What cannot travel, and what travels in unintentional ways? By focusing on the flows of material we elucidate the moral and social work that goes into the exchange of research materials and illustrate the divergences between this kind of work and the official ethics policies and frameworks that are supposed to guide it.

Our research site is a Danish laboratory that has longstanding experience of genetic research with international partners. We focus on two types of international collaboration. The first is a long-term collaboration with a research center in Pakistan aimed at studying rare autosomal recessive diseases. These diseases are more common in contexts with a tradition of consanguineous marriages. The manifest genetic disorders provide researchers with an opportunity to identify new disease-causing genes of wider relevance for understanding human biology. The second type of collaboration is an initiative taken by the Danish laboratory to establish a research consortium called The

[1] This chapter is a reworked version of a paper published in 2017 as "Ethics Policies and Ethics Work in Cross-national Genetic Research and Data Sharing: Flows, Nonflows, and Overflows" (2017) 42(3) Science, Technology and Human Values, 381–404. We thank Sage for allowing this reproduction.

International Breakpoint Mapping Consortium (IBMC). The IBMC seeks to create a saturated map of balanced chromosomal rearrangements as a way to gain functional knowledge of the human genome. Most of these chromosomal rearrangements are rare; some are unbalanced, with missing or extra genetic material causing chromosomal disorders; others are so-called balanced rearrangements in which parts of chromosomes are moved or inverted, but in a way that does not necessarily cause any pathology (see Figure 1.1). Still, the chromosomal rearrangements may cause disease if the associated breakpoints truncate a gene or an area regulating gene expression. The goal is to compile a library of these chromosomal breakpoints from which new aspects of human genetics can be investigated. The library will serve as a biomedical platform,[2] allowing easier exploration of the role of rare chromosomal rearrangements in human disease. Diagnostic and research labs from more than 50 countries on six continents are participating in the endeavor by sending samples to the laboratory in Copenhagen where they are sequenced and added to the library. The lab draws upon its long experience with international collaborations when initiating new partnerships for IBMC.

Recent years have seen great emphasis on promoting international collaborations in genetic research. Policies are developed to promote data sharing, harmonization of international rules, and open access, and global alliances are formed.[3] Mostly this work focuses on removal of so-called barriers to access, for example, by ensuring comparable demands of informed consent in different jurisdictions,[4]

---

[2] Peter Keating and Alberto Cambrosio, *Biomedical Platforms: Realigning the Normal and the Pathological in Late-Twentieth-Century Medicine* (MIT Press 2003).

[3] P. Arzberger, P. Schroeder, A. Beaulieu, G. Bowker, L. Laaksonen, D. Moorman, P. Uhlir, and P. Wouters, "Promoting Access to Public Research Data for Scientific, Economic, and Social Development" (2004) 3 Data Science Journal, 135–52; F. Colledge, B. Elger, and H. Howard, "A Review of the Barriers to Sharing in Biobanking" (2013) 11 Biopreservation and Biobanking (ahead of print); E. Dove, A-M. Tassé, and B.M. Knoppers, "What Are Some of the ELSI Challenges of International Collaborations Involving Biobanks, Global Sample Collection, and Genomic Data Sharing and How Should They Be Addressed?" (2014) 12 Biopreservation and Biobanking, 363–4. See, for example, the work of international networks such as Public Population Project in Genomics and Society (P3G), GenomEUtwin, and PHOEBE, and policy papers from, for example, the OECD. Note also the decision reached by the EU on May 27, 2016, to work towards a paradigm of "Open Science."

[4] Barbara J. Evans and Eric M. Meslin, "Encouraging Translational Research through Harmonization of FDA And Common Rule Informed Consent Requirements For Research With Banked Specimens" (2006) 27 Journal of Legal Medicine, 119–66.

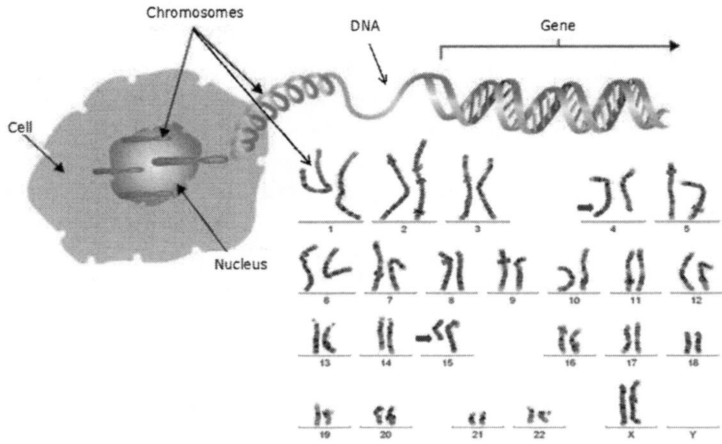

Notes: Left: The nucleus of a human cell contains 46 chromosomes, each of which is made of one long DNA double helix that contain hundreds to thousands of genes. Right: Balanced chromosomal rearrangements can be identified by examining the chromosomes in a microscope. They contain the correct amount of DNA and are in most cases considered harmless for the carrier. The image shows a balanced translocation between chromosomes 4 and 15. The translocation breakpoints are illustrated with arrows. Karyotype: 46,XX,t(4;15)(q21.3;q13).

Images with permission from NHS National Genetics and Genomics Education Centre.

Figure 1.1 Chromosomal rearrangements

shared rules of feedback of incidental findings,[5] or principles of data sharing.[6] Simultaneously, another set of policies has focused on ethical issues relating to the protection of autonomy and privacy, as seen for example in the World Medical Association's declaration on health data and biobanking. Such policies seek to control and restrict data flows. The two sets of policymaking generally develop independently, and tensions between them must be handled on the spot by those subject to their sometimes incompatible rules.

[5] Alessandro Blasimme et al, "Disclosing Results to Genomic Research Participants: Differences that Matter" (2012) 12 The American Journal of Bioethics, 20–2.

[6] Deborah Mascalzoni et al, "International Charter of Principles for Sharing Bio-Specimens and Data" (2016) 24 European Journal of Human Genetics, 1–8.

Some scholars have pointed out how data-sharing policies have been accused of neglecting the tension between data sharing and privacy protection.[7] Others have focused on how network structures involve challenges to governance,[8] as well as to funding.[9] Based on a study of networks that were claimed to be successful, Mayrhofer and Prainsack have argued that international rules are less important than informal networks in creating harmonized scientific standards and that the scientific collaboration is coproduced with ethics.[10] Some studies indicate that it is often surprisingly difficult to work across borders unless you already know the local collaborators.[11] Other strains of scholarship have articulated critiques of the many attempts to stimulate international collaboration. For example, international tissue exchange has been criticized for enacting a form of biopiracy wherein resources in low-income countries are made available without a fair return and where "benefit sharing policies" do little to alleviate the problems.[12] It has been pointed out that exchanges generate shifts in valuation of the tissue in the course of passing from research participants to local researchers and further on to international research partners,[13] and such shifts have been problematized as involving "commodification"[14] and exploitation.[15]

---

[7] Jane Kaye, "The Tension between Data Sharing and the Protection of Privacy in Genomics Research" (2012) 13 Annual Review of Genomics and Human Genetics, 415–31.

[8] I. Meijer, J. Molas-Gallart, and P. Mattsson, "Networked Research Infrastructures and Their Governance: The Case of Biobanking" (2012) 39 Science and Public Policy, 491–9.

[9] R. Jean Cadigan et al, "Neglected Ethical Issues in Biobank Management: Results from a U.S. Study" (2013) 9 Life Sciences, Society and Policy, 1–13.

[10] Michaela T. Mayrhofer and Barbara Prainsack, "Being a Member of the Club: The Transnational (Self-) Governance of Networks of Biobanks" (2009) 12 International Journal of Risk Assessment and Management, 64–81.

[11] Roger Bjugn and others, "What Are Some of the ELSI Challenges of International Collaborations Involving Biobanks, Global Sample Collection, and Genomic Data Sharing and How Should They Be Addressed?" (2015) 13 Biopreservation and Biobanking, 70–1.

[12] Cori Hayden, "Taking as Giving: Bioscience, Exchange, and the Politics of Benefit-Sharing" (2007) 37 Social Studies of Science, 729–58.

[13] Warwick Anderson, *The Collectors of Lost Souls: Turning Kuru Scientists into Whitemen* (Johns Hopkins University Press 2008).

[14] Hilary Rose, "From Hype to Mothballs in Four Years: Troubles in the Development of Large-Scale DNA Biobanks in Europe" (2006) 9 Public Health Genomics, 184–9.

[15] See discussion in Melinda Cooper and Cathy Waldby, *Clinical Labor: Tissue Donors and Research Subjects in the Global Bioeconomy* (Duke University Press 2014); Hilary Cunningham, "Colonial Encounters in Postcolonial Contexts" (1998) 18 Critique of Anthropology, 205–33; Emma Kowal, Joanna Radin, and Jenny Reardon,

In this chapter, we are interested in the work that goes into making international collaborations operational in practice and the flows of information and biomaterial which this work facilitates or obstructs. Our study of collaborations is itself the outcome of collaboration between geneticists and social scientists and an instance of data sharing. After a note on methods, we describe "infrastructures for flows" and "ethics work" to clarify how we analyze genetic research collaborations. We then examine the practical everyday type of ethics work that makes international collaborations—and thereby Big Data—possible. More specifically, we wish to explore the everyday ethics work facilitating exchange by describing what is made to flow, the nonflows, and the overflows of communication and samples.

2 METHODOLOGICAL REFLECTIONS: HOW TO EXPLORE FLOWS?

Where should you go to explore the internal workings of collaborations among research participants, researchers, and their research partners?[16] This collaboration between a geneticist (Malene Bøgehus Rasmussen) and two social scientists (Klaus Hoeyer and Aaro Tupasela) is the product of an interdisciplinary research program funded by the University of Copenhagen intended to investigate legal, ethical, and scientific challenges in crossborder sharing of biological material. The empirical data for our analysis are drawn from two interrelated research projects. The first is a genetic research project initiated by Professor Niels Tommerup to reexamine carriers of balanced chromosomal rearrangements conducted by Rasmussen (henceforth called the genetic study); the second is a social scientific study conducted by Hoeyer and Tupasela focusing on research participants, researchers, and international research collaborators affiliated with the genetic lab (henceforth called the social scientific study).[17] The social scientific study consisted of observation

"Indigenous Body Parts, Mutating Temporalities, and the Half-Lives of Postcolonial Technoscience" (2013) 43 Social Studies of Science, 465–83; Jonathan Marks, "'We're Going to Tell These People Who They Really Are': Science and Relatedness," in Sarah Franklin and Susan McKinnon (eds), *Relative Values: Reconfiguring Kinship Studies* (Duke University Press 2001), 355–83; Jenny Reardon, *Race to the Finish: Identity and Governance in an Age of Genomics* (Princeton University Press 2005).

[16] Bart Penders et al, "When Scientists, Scholars, Clinicians, Physicians and Patients Meet," in Bart Penders et al, *Collaboration across Health Research and Medical Care* (Ashgate Publishing Ltd 2015), 312.

[17] Studies of interdisciplinary collaboration suggest that it takes work and time to establish collaborations across the disciplinary boundaries of laboratory science and social science: *see* M. Albert, S. Laberge, and B. Hodges, "Who Wants to Collaborate with Social Scientists? Biomedical and Clinical Scientists' Perceptions of Social

of, and semistructured interviews with, researchers in the lab in Copenhagen (Tupasela), as well as semistructured interviews conducted during 2014 in Denmark (Hoeyer) and Finland (Tupasela) with Danish research participants (Hoeyer) and Pakistani (Hoeyer) and Finnish (Tupasela) collaborators. The interview framework explored the hopes and concerns that research participants, researchers, and research collaborators may or may not attach to the use of the samples and related information as they move across the globe. Quotes are translated from Danish and Finnish by the authors.[18]

Research participants' views are primarily explored through interviews with Danish individuals participating in the genetic study. The interviewees were recruited through the genetic project and the interviews were carried out by Hoeyer.[19] Adult carriers of balanced chromosomal rearrangements who lived in Denmark were invited to participate in the genetic study, which included a questionnaire on health-related issues. One question asked participants if they were interested in being contacted about participating in the social scientific study to offer their opinions about the research. Research participants for the social scientific study were chosen by the authors to maximize variation along classic demographic variables such as age, occupation, education level, and place of residence, as well as differences in response to the invitation to participate in the original genetic study. Enrolment of participants in the genetic and the social scientific projects followed two separate research protocols approved by the Danish Data Protection Agency.[20] As part of the genetic study, Rasmussen had been in contact with several research participants and knew that some participants had responded with enthusiasm, some did not seem to pay much attention to it, and others had expressed anxiety and concern. Such differences in attitude toward the genetic study also were

---

Science," in Bart Penders et al, *Collaboration across Health Research and Medical Care* (Ashgate Publishing Limited 2015), 59–80. We, too, had to gradually negotiate the terms of our own collaboration and establish relationships to make meaningful the sharing of facilities and research network in Denmark, Finland, and Pakistan. In this way, our own collaboration came to mirror the phenomenon that the social scientific study had set out to describe, namely how flows of data are established through social work, though we only gradually realized this point in the course of writing the chapter.

[18] We would like to thank Zainab Sheikh for helping with the Urdu translation.

[19] The genetic study was approved by the Danish Data Protection Agency (J.nr. 2012-54-0053) and by the Danish regional Research Ethics Committees (J.nr. H-KF-2006-5901). Health-related information obtained through the genetic study was never disclosed to or discussed within the forum of the social scientific study, in agreement with the Danish Data Protection protocol (J.nr. 2012-54-0053).

[20] J.nr. 2014-41-3055: Global Genes, Local Concerns: Legal, Ethical and Scientific Challenges in Cross-National Biobanking and Translational Exploitation.

used to select potential candidates for semi-structured interviews.[21] A total of 32 participants from the genetic study were invited to engage in the social scientific study. Two opted out, one did not reply, and six engaged only in email-based discussions, while 23 were interviewed either in their homes, in the office of Hoeyer, or over the phone, depending on their personal preference and logistical opportunities. The interviews lasted between 15 minutes and almost 3 hours. Most took about an hour. After approximately 15 interviews, we began reaching data saturation in the sense that we heard similar hopes and concerns repeated. All interviews were recorded and transcribed.

Interview candidates among the Danish researchers and their collaborators were identified using the snowball method, as well as through the identification of key partners in the IBMC. Beginning with the head of the genetic lab and a key staff member, we moved on to research partners in Finland and Pakistan, and some of their staff members. The Pakistani researchers either worked in Copenhagen or came on regular visits, while the Finnish partners were interviewed in Finland. We were only able to do interviews with Danish research participants, because we were able to ask for their permission through the original enrolment questionnaire used for the genetic study. The rules set by the Danish Data Protection Agency and Research Ethics Committee system have shaped what we could do with this analysis, and so understanding the rules became an interesting parallel study that also informed the analysis. While we have collaborated to understand how we might enhance the ethics of research collaborations, the ethics policies that are expected to guide collaborations have often felt like arbitrary restrictions. Rather than stimulating consideration of how to respect the autonomy and integrity of the interviewees, we have often found ourselves speaking about what we were *allowed* to do. Ethics rules, by way of being "rules," have thus stimulated conversations about legality rather than values; about limits to what we could do, rather than thoughts about what we should do. We have used this observation as an impetus to reflect on the ethos of our own work and how it relates to the rules that are supposed to guide it and ensure its legitimacy.

[21] To ensure compliance with rules set by the Data Protection Agency, neither Hoeyer nor Tupasela was informed about who fit which criteria. Similarly, the anonymity of the interviewee was upheld the other way round. Interviewees were informed that the researchers from the genetic study would not know who said what to Hoeyer and that their participation in the genetic study would not be affected by their decision to give interviews. As a member of the genetic study, Rasmussen has not seen the individual transcripts and was only informed about interview contents in anonymized form.

3 DEFINING ETHICS WORK AND INFRASTRUCTURAL FLOWS AS ANALYTICAL OBJECTS

By focusing on flows we are inspired more broadly by what Sheller and Urry have described as a "mobility turn" in the social sciences.[22] The mobility turn has involved paying increased attention to what moves, how it moves, and what are seen as legitimate and illegitimate movements. Flows of samples and information are coproduced with an infrastructure facilitating the movement. Following Star and Ruhleder's seminal paper on research infrastructures,[23] we think of infrastructures as activities rather than things. Everyone and everything is simultaneously engaged in many different relations, hence there is not *one* infrastructure, but multiple interrelated infrastructures enacted through different practices. The Danish laboratory of course still depends on a material infrastructure of freezers and information software, but such hardware is simultaneously engaged in many other flows.

A flow of material depends on much more than freezers.[24] Groundwork is required for people to donate samples and for researchers to ship the samples to Copenhagen. We think of this work, aimed at enacting sustainable relations, as "ethics work." As argued by Mayrhofer and Prainsack,[25] the scientific collaboration is coproduced with ethics through informal networks. Still, the practical and mundane ethics work that is necessary to make material and information flow continues to receive limited attention. With "ethics work" we build on the social study of medical ethics as a practical activity, not just abstract values or principles.[26] This involves focusing on what people do, rather than what they think they ought to do. In the following, we wish to focus on the otherwise tacit aspects of this work, because we believe it is central to

---

[22] Mimi Sheller and John Urry, "The New Mobilities Paradigm" (2006) 38 Environment and Planning A, 207–26; John Urry, *Sociology beyond Societies: Mobilities for the Twenty-First Century* (Routledge 2000).

[23] Susan Leigh Star and Karen Ruhleder, "Steps toward an Ecology of Infrastructure: Design and Access for Large Information Spaces" (1996) 7 Information Systems Research, 111–34.

[24] S Leonelli, "What Difference Does Quantity Make? On the Epistemology of Big Data in Biology" (2014) 1 Big Data & Society, 1–11.

[25] See n 9.

[26] Charles L. Bosk, *Forgive and Remember: Managing Medical Failure* (University of Chicago Press 1979); Daniel F. Chambliss, *Beyond Caring: Hospitals, Nurses and the Social Organization of Ethics* (Chicago University Press 1996); P. Wenzel Geissler et al, "'He Is Now Like a Brother, I Can Even Give Him Some Blood'—Relational Ethics and Material Exchanges in a Malaria Vaccine 'Trial Community' in The Gambia" (2008) 67 Social Science & Medicine, 696–707.

international research collaborations. Whereas Sleeboom-Faulkner and Patra emphasize the exploitive potential of international collaborations that bridge and even thrive upon inequalities,[27] we focus here on what people do to engage the exchanges. We wish to elucidate the tacit work done to respect the involved stakeholders, because we believe it should be acknowledged as central to good and sustainable research ethical practices.

How should we conceptualize the objects around which collaborations revolve? Are we dealing with "*a* biobank," for example? The notion of *a* biobank is problematic because samples used in international collaborations often belong to many different "banks." Samples derive from numerous different sources, such as hospitals and research projects, and they are used for very different purposes. They form part of very diverse flows. In some cases the samples are not even stored at the lab: they are completely used up in the course of the research, leaving only data behind. The concept of "biobank" might be both too singular (one set of samples in one place), too static to capture the sense of flow (it indicates accumulation), and too informed by one type of purpose (research) to capture all the involved flows and uses. Would the concept of platform then work as an alternative to biobank? Indeed, Keating and Cambrosio suggested the concept of the biomedical platform to tie in the social and material aspects of networks in ways that simultaneously captured the agential capabilities of such networks.[28] However, if you ask around in the Copenhagen lab, the word platform figures in a quite specific sense: it signals, for example, the equipment used for sequencing (for example, a technician explained: "The Illumina platform can sequence 32 mate-pair samples in one 11-day run"). Other technical "platforms" are used for other purposes, such as "genome browsers" to look up what is already known about a particular genetic sequence. The IBMC map is expected to serve a similar function. The concept of platform therefore already holds a slightly different meaning, and it poorly captures the range of social activities and all the tacit work through which flows are enacted in international collaborations. We therefore suggest focusing on what collaborations aim to do—enact *flows* of biological material and data—and the work that goes into making it happen. With a focus on flows we deliberately bring back in the human actors, and the stories that matter to them.

We begin by describing the practical work going into the creation of a flow of genetic material and probe the hopes and concerns that travel along with

---

27 Margaret Sleeboom-Faulkner and Prasanna Kumar Patra, "Experimental Stem Cell Therapy: Biohierarchies and Bionetworking in Japan and India" (2011) 41 Social Studies of Science, 645–66.

28 See n 1.

it. To do so, we first reflect on the different strategies for identification of research participants and then provide examples of how participants are enrolled in Denmark, Finland, and Pakistan.

4 FLOWS: WHAT FLOWS AND WHAT MAKES IT FLOW?

The Danish laboratory recruits research participants in different ways depending on the options available in each country. In Denmark, potential genetic research participants (that is, carriers of chromosomal rearrangements) are identified through the Danish Cytogenetic Central Register.[29] In Finland there is a similar register, but it is run by a researcher and all access to the register goes through this person. In Pakistan, recruitment of research participants is based more on a snowball method and happenstance, since there are no centralized registers or databases for rare diseases. The Pakistani samples are not currently used for the IBMC map because these samples represent different kinds of genetic disorders. But the mode of collaboration has informed the work with the IBMC map, and the samples collected are used toward the same overall purpose of understanding genetic causes of disease.

To begin the genetic study, the lab in Denmark first seeks approval from a Research Ethics Committee and the Data Protection Agency. It then receives a list of carriers of chromosomal rearrangements from the Danish Cytogenetic Central Register. However, the lab does not approach the potential research participants right away. First, each diagnostic facility having entered patients into the register is approached and an agreement is made with them on how to approach potential research participants. They need to determine whether the carriers are aware of their carrier status, as for ethical reasons they do not wish to contact people who have not previously been informed of their status. This information is gathered from the original medical record describing the chromosome analysis performed—a point to which we return in the next section. The participants are invited to participate through a letter containing information describing the project, a consent form, and a health-related questionnaire. The participants are also informed that they might be invited to donate a biological sample such as a blood sample. The lab keeps records of all formal aspects of the research project as covered by the research protocols for the Research Ethics Committee and the Data Protection Agency, including consent status of participants, whether participants wish to be informed about clinically relevant results from the project, and if a biological sample has been

---

[29] Johannes Nielsen, *The Danish Cytogenetic Central Register: Organization and Results* (Thieme-Stratton 1980).

donated to the project. This all follows official guidelines and international standards. In addition, however, the lab finds it necessary to record questions or concerns relating to the project articulated by the research participant in order to ensure that more personal concerns are respected. It might be that a particular family member is not informed about his or her carrier status, and therefore should not be approached by the research team. In this way, some hopes and concerns travel along with samples and health data, though they flow through separate record systems.[30]

What do the Danish research participants think about such records of social and moral concern? In our interviews, some participants expressed very strong concerns about particular family members who should not be informed about the identified chromosomal rearrangements. For them, keeping records of those concerns was essential to their trust in the researchers, though it was not conceptualized as a particular task demanding its own records. Researchers are just expected to *know*. One man had decided not to tell one of his children about their shared carrier status, and when he received the invitation to participate in research he was cited as thinking: "What then might happen tomorrow? … How about my grandchildren; I don't know, are they also gonna be summoned? Well, then I need to inform [the child] first … there is an element of fear in all of this." He relied on the researchers to realize that only some family members were aware of their carrier status. Interestingly, it was more common for the research participants to state that they were happy to participate in genetic research as long as researchers accessed *only* biomedically relevant information, and knew nothing about their personal hopes and concerns. One woman, for example, said that she did not want the researcher to know anything about her choices and preferences, but they could use her blood for whatever they wanted. She continued: "I think technologies have progressed very far, but they probably cannot figure out anything *personal* about me by looking at my blood … So they can do whatever investigations that they want on my old blood. I don't have any problem with that." Ironically, for these research participants, keeping records of their stated hopes and concerns, which is meant to

---

30 Terms of participation in biomedical research projects are covered by guidelines from the Research Ethics Committee system and the Danish Data Protection Agency. These agencies prescribe the formal requirements for informed consent, including information about research aims; benefits and risks of participation; terms of use of biological samples, including storage and limitations of use; confidentiality; and voluntariness. However, the Research Ethics Committee system and the Danish Data Protection Agency provide only limited instructions on how to address and keep track of more subtle issues such as the moral concerns uttered by the research participants. Recording what people say has become a tacit form of ethics work emerging out of practical experience and moral engagement with research participants over many years.

protect their privacy and respect their autonomy, is aligned with a precarious sense of concern, while genetic information and health-related data are seen as less problematic. A record of moral concerns and participant choices is nevertheless instrumental for respecting their position. Keeping records of personal choices is part of demarcating what should and should not flow, but these records themselves are not supposed to flow anywhere.

In Finland, the register for balanced chromosomal rearrangements is run and managed by one researcher who has visited all the hospital regions that have a genetic counseling and/or analysis unit. At these units, he has collected all the information available about individuals and families who have been identified with chromosomal rearrangements and entered them into a register, which his group uses for research. The register is not part of the national infrastructure but is run by this researcher. Once identified and entered into the register, the patients were first sent a letter regarding the research and given the possibility of contacting and discussing their problems with a genetic counselor or other senior physician. Later, if they consented to participate in the research, they were sent a sample collection kit that they could take to their local clinic to have a sample taken, which was then returned to the researcher. This collection kit included the contact details of a research group member who could be contacted if there were questions regarding the sampling procedure. During the process of collecting samples from about 100 patients who had requested the sample kit, one of the research group members noted that she was contacted by some patients who wanted to tell her their "stories" and experiences of living with their condition. In many cases, this related to having had a miscarriage, but other medical conditions were also discussed. Patients apparently wanted to share their experiences with the researcher in a way that was not sufficiently represented or covered by the sample collection or medical history approach. Though this information needs to flow for the patients to feel confident, it does not travel along with the samples to Copenhagen, as we shall discuss in the section on nonflows below. Some patients were also concerned about how information produced in the research would be shared, since they did not want information to be shared with family members. The flows were conditioned on nonflows. Other patients wanted additional information regarding chromosomal rearrangements and the research itself. In such cases, the flow of genetic material in one direction rested on genetic knowledge flowing in the opposite direction to the research participants. Again, social skills are needed to create genetic material flows: infrastructures depend on important, albeit tacit, forms of ethics work which do not feature in the official documents of collaboration.

The recruitment process in Pakistan is as dependent on social skills, or what we call "ethics work," as it is in Denmark and Finland—but in significantly different ways. Samples are typically collected from people who do not have access to any of the genetic counseling services through which the Danish and

Finnish participants are identified. As mentioned previously, there is a high prevalence of genetic disorders, which is seen by scientists as related to high rates of intrafamily marriages. Scientists explain how, to gather samples and medical information, they first have to establish a relationship with these families. They do so through research students who are recruited to work for the institute, partly based on where they come from: by having students from a wide range of local areas, the institution has a network of people familiar with relevant local knowledge. Once contact is established through the local student, the institution can offer various forms of genetic counseling services, and in their interviews the scientists describe how they sometimes bring family members from rural areas to hospitals in major cities where they can undergo more sophisticated diagnostics and treatment offers.[31] Diagnostic, treatment, and research activities are thus fully intertwined, though certain bioethical ideals tend to suggest that they are better kept safely apart.[32] All these activities have to take place in order to create the relationships needed for information and samples to begin their travel.[33] When samples travel on to Copenhagen, it is often in the company of PhD students or postdoctoral researchers undertaking training in Denmark. These researchers bring with them an awareness of the stories of the people in whom the samples originated, and they are concerned about the ability to return information to the patients and their families.

In many ways, the samples traveling across the globe thus remain attached to the hopes and concerns through which they were first produced. However, some samples are sent without stewards, and samples are used for research that the Pakistani researchers are not themselves directly involved in executing. The international collaboration here relies on trust between the research partners (rather than between researcher and research participant). When characterizing this trust, one of the researchers used the word *Bharossa* [بھروسہ]: "It is, you could say, a super degree of trust, Bharossa [بھروسہ] ... When you rely on somebody that you have the highest level of trust in, then it is Bharossa!" Bharossa typically belongs to a religious idiom signaling faith, and the researcher thus emphasizes a very different kind of social contract than what can be contained in Material Transfer Agreements (MTAs) or

---

31 Prasanna Kumar Patra and Margaret Sleeboom-Faulkner, "Informed Consent and Benefit Sharing in Genetic Research and Biobanking in India: Some Common Impediments in Practice," in Peter Dabrock et al (eds), *Trust in Biobanking* (Springer 2012), 237–56.

32 Paul S. Appelbaum et al, "False Hopes and Best Data: Consent to Research and the Therapeutic Misconception" (1987) 17 The Hastings Center Report, 20–4.

33 See also Anderson, n 12; Emma Kowal, "Orphan DNA: Indigenous Samples, Ethical Biovalue and Postcolonial Science" (2013) 43 Social Studies of Science, 577–97.

policies of open access. Hoeyer asked the researcher what he would consider a *breach* of trust. He replied: "If they send the sample to a third party without my knowledge, which they have never done. Whenever they correspond, our collaborator, with a third, fourth, or even fifth research group, they always ask me." Samples are made to travel in a sociomaterial infrastructure that involves much more than data storage and exchange. Notions of open access as a plain matter of sharing everything with anyone seem to ignore how collaborations are socially embedded. There is a morality built into the ties between people, and these ties restrict the flows. This takes us to the nonflows of research collaborations.

5 NONFLOWS: WHAT CANNOT FLOW AND WHICH FLOWS ARE STOPPED?

In addition to the formally regulated types of nonflows, such as ensuring confidentiality by not circulating patient names or social security numbers among researchers, there are other, more informal aspects of daily activities that stop material and information from flowing. In this section we look at nonflows, first by looking at how the creation of social ties between research partners can prevent them from entering other partnerships. We then examine examples of nonflows of information in the form of what can be termed "strategic ignorance,"[34] where some aspects of research collaboration are not revealed to research participants because they are expected to dislike them.

Concerning mutual obligations that imply nonflows, a Pakistani researcher noted that he was "morally obliged" to continue working with the Danish lab, and therefore unable to initiate collaborations with others when these would collide with the research interests of the Danish lab. Similarly, the diagnostic labs delivering samples for the IBMC map are expected not to support competing attempts of breakpoint mapping, although there are no formal rules against that. So the establishment of a flow in one direction creates interdependencies and limits to other flows. We might say that "trust" implies an outside border delineating what must be excluded from the relationship. It is well known that researchers often try to ensure a lead position by way of controlling resources, while both patient organizations,[35] on the one hand, and funding agencies and authorities (such as the Organisation for Economic Co-Operation and

[34] Linsey McGoey, "Strategic Unknowns: Towards a Sociology of Ignorance" (2012) 41 Economy and Society, 1–16.

[35] Georg Lauss, "Sharing Orphan Genes: Governing A European-Biobank-Network For The Rare Disease Community," in Peter Dabrock et al (eds), *Trust in Biobanking* (Springer 2012), 219–35.

Development (OECD)),[36] on the other, seek to foster wider sharing of samples. It can be interpreted as a simple conflict of interest. However, what we suggest here is that some nonflows might not be about ensuring an edge in terms of competition at all. We might have to acknowledge that relationships of trust that are stable enough to facilitate flows of sensitive material cannot easily be made "open access."

The notion of an outside border is also embedded in rules of *confidentiality* as they relate to research participants, irrespective of the national context in which the collection takes place. As a consequence, the records of personal histories and concerns that each lab keeps do not flow. When, for example, samples are sent from Finland, they are coded and contain no information on the person from whom they were collected. Even when Pakistani researchers travel to Copenhagen and work on samples they have collected themselves, they rarely share the personal histories of the participants with the lab technicians or other research partners. At some point, the samples travel unaccompanied by social hopes and concerns, partly because rules of confidentiality ensure that nothing personal is communicated. In this way, ethics rules can limit the continued awareness of local moral concerns. The lab in Copenhagen collaborates with another lab in which scientists do experiments on zebrafish to explore functional capacities of the identified chromosomal rearrangements. Illustrating how the samples become stripped of individual histories, a staff member in this biology lab stated in a conversation with us that he never considered the people behind the samples. Thus, the collaborative research is produced through social skills, as illustrated previously, but it simultaneously disentangles genetic information from social histories.

Finally, there are elements of the sample collection that involve what is described in the literature as strategic ignorance: planned nonflows of information.[37] In order to operate in rural areas of Pakistan, researchers have to take into account how their actions can be interpreted and thus avoid sharing particular forms of information. Following the capture of Osama bin Laden, and public statements about his identification through the collection of blood samples

---

[36] Organisation for Economic Co-Operation and Development (OECD) "OECD Principles and Guidelines for Access to Research Data from Public Funding" (2007), www.oecd.org/science/sci-tech/38500813.pdf (accessed 25 June 2018).

[37] Paul Wenzel Geissler, "Public Secrets in Public Health: Knowing Not to Know while Making Scientific Knowledge" (2013) 40 American Ethnologist, 13–34; Linsey McGoey, "Strategic Unknowns: Towards a Sociology of Ignorance" (2012) 41 Economy and Society, 1–16; Robert N. Proctor, "Agnotology: A Missing Term to Describe the Cultural Production of Ignorance (and Its Study)," in Robert N. Proctor and Londa Schiebinger (eds), *Agnotology: The Making and Unmaking of Ignorance* (Stanford University Press 2008), 1–36.

from his son during a polio vaccine campaign, the collection of samples by outside agencies in that area has been surrounded with anxiety, questions, and doubts. One of the researchers taking part in the collection explained: "If you go to that area, and any member of the team is not known by the people where we go, then they ask 'Who are you, what is your objective, what will you do, will you give this blood to the CIA or what is the purpose of this?'" Collaboration with a Western lab does not help build trust in this environment. Furthermore, Denmark is discussed as a problematic country following the publication there of a series of cartoons depicting the prophet Muhammad. As a consequence, sometimes field staff decide to downplay the international aspects of their science collaborations when explaining their research to local participants, or they provide explanations only after a proper relationship has been established. Then the thought of samples traveling to Denmark can in fact be comforting, a Pakistani researcher explained, because it is taken as proof that the researchers take their condition seriously. Again, this initial nonflow is not meant to deprive the research participants of an "autonomous choice." Rather, it is seen by the researchers as a way of expressing care for local sentiment in order to enroll them into research that can hopefully help them: giving them information about the condition and their potential carrier status can be essential for family planning. Information is not seen as a one-off thing that should be given prior to all involvement, as some consent policies suggest. Informational needs are seen as developing along with the relationship. The flows of information that eventually make up the infrastructure for the genetic research are intertwined with nonflows, protective boundaries, outside borders, and silences. Such nonflows are as socially embedded as fully articulated hopes and concerns.

The flows (and nonflows) we have described produce genetic knowledge that can operate free from the specific social histories of the research participants. For the most part, researchers and lab technicians working with samples in the lab consider their work routine and avoid attributing any personal attachment to them. However, it is important to note that there are several cases of overflows, where unintended information and unwarranted hopes and concerns travel along with the samples into the labs and from the labs into the homes of the people donating the samples. It is to such overflows that we now turn.

6 OVERFLOWS: WHAT TRAVELS UNINTENTIONALLY?

Though open access and data sharing policies intend to provide access only to scientifically relevant material, some forms of social meaning might also flow along with them. Several lab technicians working with diagnostic samples, for example, remarked that it was difficult not to consider that a positive test result could have significant implications for the affected family. The fates of the unknown people taking the diagnostic tests thus come to matter even when the identities of the persons remain unknown. In such situations, genetic information overflows with social concerns.

More significant, perhaps, are the types of overflow that move in the other direction: from the research activities to the research participants. One of the Pakistani researchers described how collection of samples could interact with negotiations of etiology. When collecting samples, she found herself negotiating far more than the terms of participation in research: she would find herself introducing novel conceptions of the causes of illness that move the understanding of, for example, schizophrenia from a religious realm into a secular biological realm. The research activity thus leaves behind new narratives of illness in the involved families. In other cases, narratives of genetic illness can cause a fear of stigma: irrespective of the scientific explanations given, some families experience reactions from others who think that it is better to avoid marital relations with particular strains of the family tree. The global networks of science will always interact with these very personal stories in unwarranted ways. Science is unruly and not subject to total legal, ethical, or scientific control. Its material infrastructure overflows with meaning.

One might expect the mountainous regions of Pakistan to harness greater interpretive flexibility with respect to the scientific narratives than is the case among the literate Danish population. However, local interpretations are a consistent feature of the stories told also by Danish research participants. Several Danish research participants, for example, explained how the questionnaire they filled in as part of their participation in the genetic study led them to associate their chromosomal rearrangement with a range of personal characteristics. Some said they had come to feel more closely related to other family members with the same rearrangement. Others describe how they used the questionnaire as a source of information about "what science now knows," and used it to understand the cause of particular diseases among family members, for example. The research activity thereby sets in motion narratives about health and illness that are founded not on science but on local interpretations by people looking for explanations. People use pieces of information picked up during the research activity to symbolically make sense of their world.[38]

There are also people with balanced chromosomal rearrangements who wonder whether they can be organ or blood donors or whether their genetic

---

[38] Klaus Hoeyer, "Traveling Questions: Uncertainty and Nonknowledge as Vehicles of Translation in Genetic Research Participation" (2016) 35 New Genetics and Society, 351–71.

makeup might cause danger to potential recipients. One man, for example, had registered an optout for organ donation despite actually supporting organ donation, because he feared his rearrangement could cause an organ rejection. In the course of an interview he said: "And then some poor fellow, who is already very ill, gets my organ, 'hurrah!' And then two weeks later 'sorry, mate, it didn't work out, it was rejected'. I would feel like a villain [slyngelagtig]." Following the interview, Hoeyer informed him that he had little need to worry about such a rejection; after getting confirmation from Rasmussen based on an anonymized description of the case, Hoeyer was able to further console the man, who then wrote back that he would change his decision to allow organ donation. Again, such overflows emerge as people apply the information acquired through research participation to concerns of their own that are not covered by official guidelines. And again, the research participant was consoled thanks to our collaboration, and despite—rather than because of—the official guidelines, which were hindering passage of information from the social scientific study to the genetic study.

Researchers in the genetic study are keenly aware of potential overflows, but they are not in a position to control them. Importantly, the official ethics policies focus on what people *should* know *prior* to accepting the invitation to research, not on how to work with such unintended *consequences* of research participation. Such work demands funding and continued dedication. It is a very different kind of work than that involved in making data available in open access depositories, but if the moral concerns of donors count it cannot be neglected. Researchers, research participants, and funding agencies may operate with very different definitions of what constitutes appropriate sharing and what constitutes appropriate flows, preferred nonflows, and overflows. It largely remains a personal query for researchers to contemplate these issues and to figure out how to respond to them.

7 CONCLUSION

We have described how recent years have seen a substantial policy emphasis on stimulating international collaboration in genetic research through data sharing policies and global regulatory harmonization. We have argued that when policymakers think of the task as one of encouraging researchers to "share" and "remove barriers," they might misconstrue research collaboration as a simple matter of providing access to preexisting freezers and databases. Data sharing does not materialize without work, however, and this work is socially embedded. To promote data sharing, it is important to understand the social mechanisms involved. Networks that support the flow of sensitive material and information may not easily enable "open access" because the social mechanisms that facilitate the flow depend on clear expectations about mutual social obligations. They build on relationships, not just infrastructures of data availability. Big Data depends on an ethics of detail. Though we have explored genetic research collaborations, similar social mechanisms also might shape research in other fields. Our findings suggest that effective social relations constitute an important prerequisite for data sharing. The dynamics of flows, nonflows, and overflows are configured in different ways, depending on the type of sharing network that is in question. Where policies of "open science"[39] tend to imagine that material and information will flow if only an infrastructure is provided (and a demand for sharing is installed), we instead suggest paying increased attention to the work going into the making of flows.

Furthermore, resistance to data sharing can reflect concerns other than self-interested attempts to ensure a competitive edge. We have argued here that, in some cases, the social ties of mutual obligation that facilitate the flow of material in the first place also install nonflows to preserve the sense of trust needed to collaborate. If the genetic biobank compiled in Copenhagen is to become a "platform" for *future* research,[40] we have shown here how it is also a materialization of *past* collaborations and acts of care heavily dependent on social skills.

We believe that it is important to acknowledge the practical kind of work that we have called "ethics work." By staying aware of the tacit ethics work, we might better acknowledge all the "emotional labour"[41] researchers need to do to make genetic material flow. Data sharing among participants and researchers does not just happen; it is made to happen. A lot of care is needed for the scientific enterprise to connect to the people in whom the samples and health data originate, and to create relations of trust between researchers.[42] The making of Big Data depends on a form of detailed attention to local concerns that grand harmonized ethics policies do not ensure.

The accumulation of samples and data cannot be limited to the compiled material, knowledge, or machinery typically associated with "data sharing," and the social skills on which flows of genetic material depend should not pass unnoticed. Such skills reflect the everyday ethos of work for researchers rather than the ethics policies intended to guide the international networks

---

39 General Secretariat of the Council, "Council Conclusions on the Transition towards an Open Science System. Brussels, Council Of The European Union" (2016 May 27, REF 2596/16).

40 See n 1.

41 Arlie Russell Hochschild, "Emotion Work, Feeling Rules, and Social Structure" (1979) 85 American Journal of Sociology, 551–75.

42 Carrie Friese, "Realizing Potential in Translational Medicine" (2013) 54(S7) Current Anthropology, 129–38; Annemarie Mol, *The Logic of Care: Health and the Problem of Patient Choice* (Routledge 2008).

through which they procure samples. The lab in Copenhagen has developed ways of keeping records of social concerns and modes of collaborating with international partners. It represents a form of ethics work that cares for detail. The emphasis in the lab's global collaborations is on the establishment of *enduring* engagements with partners and research participants. This implies returning information to those who want it, inviting international partners to work in the lab, and (in the process) creating social ties extending beyond plain data sharing. The tacit aspects of research collaborations remain largely unfunded, however, and this type of work does not provide much academic credit. If we begin acknowledging this important ethics work, we might also consider including it in new applications, discussing it in guidelines, and building future policies on experiences with this very practice-oriented form of ethics. This implies that funding agencies and research regulators should not impose demands of data sharing without paying attention to, and ensuring the necessary funding for, the work going into ensuring sustainable social relations among research participants and research partners. As such, the building of research infrastructures should focus more on fostering and nurturing social relations among actors.

The tacit forms of ethics work just described deserve dissemination to other fields seeking to implement genetic counseling based on emerging "Big Data" tools; for example, in attempts to introduce whole genome sequencing and exome sequencing into everyday clinical care.[43] The first step is to make it visible—and to make ethics debates focus on the mundane practices of care just described—with the same level of attention as has been paid to the abstract ethics principles guiding existing policy work. Only when we discuss what has already been done may we aspire to make data sharing policies and research ethics rules relevant for the science practices they are supposed to guide. In relation to the mobility turn and an increased interest in Big Data,[44] close scrutiny of actual flows will not only contribute to the study of data sharing policies; it will also form part of a larger corpus of scholarship exploring the preconditions for, and implications of, an ever more data-intensive research paradigm.

---

[43] Stefan Timmermans, "Trust in Standards: Transitioning Clinical Exome Sequencing from Bench to Bedside" (2015) 45(1) Social Studies of Science, 77–99.

[44] Sheller and Urry, n 21.

2. Biobanks as knowledge institutions

Michael J. Madison

1 INTRODUCTION

Among the earliest references to biobanks in the scholarly literature is a paper titled "BIOBANK, A Computerized Data Storage and Processing System for the Vascular Flora of Iowa," published in 1979 in the Proceedings of the Iowa Academy of Science.[1] While thoughtful and structured collections of biological specimens date back hundreds of years, to early "physic gardens" of plants cultivated for their medicinal properties,[2] that newer usage of "biobank" identifies a pair of contemporary themes. First, how are we to construct systems that effectively integrate material samples with accompanying immaterial information or data, in a usable way? Second, how are we to ensure that multiple systems of that sort are sustainable over time as matters of scientific practice and economic support, and that they are interoperable with each other? These are questions of both law and technology. Biobanks today are both critical institutions in their own right and critical illustrations and examples of broader governance questions concerning shared knowledge resources.

In that regard, this chapter makes and examines three related claims. First, biobanks are knowledge institutions, both in the sense that they store knowledge and information about the world in material and immaterial forms and in the sense that they enable researchers to produce new knowledge, including useful applications of basic knowledge. Second, the broader questions of design and management that biobanks pose are resource governance questions rather than questions solely of law or of public policy. Third, despite the varied and diverse nature of biobanks today (indeed, precisely because of their

---

1 Lawrence J. Eilers, "BIOBANK, A Computerized Data Storage and Processing System for the Vascular Flora of Iowa," (1979) Proceedings of the Iowa Academy of Science 86: 15.

2 American Medical Association, "A History of Botanic Gardens," (1915) Journal of the American Medical Association 65(2): 170; Edward M. Holmes, "Horticulture in Relation to Medicine," (1906) Journal of the Royal Horticultural Society 31: 42.

diversity), their social and scientific importance dictates the need for a robust program of research of a comparative nature to identify shared features that contribute to their success (where they succeed) and features that likely contribute to problems or even failure. Both the importance of biobanks and associated governance challenges have only grown larger and more complex as biobanks meet the era of data science. In that regard the chapter points to examples of emerging scholarly literature that focuses on governance challenges of material and data in biobank contexts. The chapter suggest directions for future work, building on the emerging knowledge commons governance research framework.

2 A SERIES OF RELATED DEFINITIONS

This section sets out some preliminary definitions.

2.1 Biobanks Defined

Biobanks today are defined generally (along with the related terms "biorepository" and "Biological Resource Center," or BRC) to include structured collections of biological materials and associated data, stored for purposes of both present and future scientific research. Materials may be collected from humans, animals, cell and bacterial cultures, and plant and other environmental resources. Repositories of human biological materials and accompanying data generally are referred to as biobanks. Institutions that deal with plant and environmental samples generally are referred to as biorepositories or BRCs. Collections vary widely in size (from population-based collections to small collections for clinical or academic study), in purpose (from disease-specific to sample-based research), and in organizational status (nonprofit to academic to commercial). Biobanks and biorepositories each raise some distinct governance concerns, and biobanks as such are often analyzed as an institutional mode in order to isolate legal and public policy concerns that are linked to storage and use of human specimens and related data.

For the purposes of this chapter, the similarities among these institutions take precedence over the distinctions. Here, biobanks include all repositories of biological information, stored in any manner and on any media, including information stored primarily as data rather than primarily as tissues or biological specimens or samples. In all cases, biobanks and biorepositories are resource pools, composed of some population of agents contributing specimens and related informational material, some population of agents (perhaps overlapping with the first) having the power to access and perhaps withdraw material, and a character defined by the fact that the social value of the pooled samples likely exceeds the value of each sample considered in isolation.

In sum, biobanks and biorepositories are resource sharing institutions, not solely exchanges or clearinghouses for access to individual samples. They are not merely tissue and specimen banks. Associated information and data may be hand-collected and curated, but increasingly they are apt to be collected and stored in high-speed and large-scale networked databases and related systems. Only a thin line may distinguish a biobank of material specimens or a network of biobanks focused on related types of specimens, on the one hand, from a large-scale bioinformatics dataset, on the other. If that distinction may be somewhat easier to describe today, then it may become more fluid in the near future. Today, biobanks are necessarily intertwined with data science.

The contemporary challenge for researchers is to grapple with the range and diversity of biobanks so that they can be analyzed in some systematic way. Despite such an inclusive beginning, or perhaps because of definitional and classification problems in the broad biobank field,[3] no single, integrated census of biobanks provides their total number.

2.2 Knowledge Defined

Biobanks exist to preserve knowledge and information in systematic ways for future generations (the preservation, curatorial, or stewardship function) and also to support generating new scientific and medical knowledge (the production function). Implicit but critical in that framing are definitions of knowledge and information. This chapter does not limit itself to formal or technical definitions of either one. "Information" is often regarded as "raw" or "unprocessed" descriptions about the natural, physical, or social world. "Knowledge" is often regarded as "refined" information, "refined" because it has been treated by human analysis and converted into something that approaches shared scientific or cultural truth. Both knowledge and information are described primarily as immaterial "things," that is, conceptual objects of human engagement and thought.

This chapter adopts a different, more expansive, and more fluid framework for describing (rather than defining) knowledge and information, in order to capture the range of functions that biobanks perform. Knowledge and information are related expressions for the proposition that intellectual, creative, and scientific investigation of human, natural, and physical experience leads to the embodiment of those investigations in shared thoughts and ideas, in

[3] Gail E. Henderson et al, "Characterizing Biobank Organizations in the U.S.: Results from a National Survey" (2013) Genome Medicine 5: 3.

conceptual objects, and in material objects.[4] Those embodiments are identified and defined primarily in pragmatic terms, because the definitions align with the functionality of the resulting objects for practical uses, rather than in strict ontological terms. Scientific knowledge remains a source of truth in the sense that it represents the shared understanding of an expert community of practitioners. The development of digital technology during the later twentieth century exposed the lack of utility in sharp distinctions between conceptual and material objects (such as ideas and things); between static forms of knowledge and dynamic processes of knowledge production, distribution, and use; and between micro and macro scales of knowledge.

Each of these perspectives contributes something valuable to our understanding of the world. In the context of biobanks, each individual specimen is a source and embodiment of information about itself; in the pool of specimens and data that composes the biobank, they are parts of a larger knowledge resource that is a source and embodiment of information about a larger collective of humans, plants, and so on. The specimen is a knowledge component; the biobank is a knowledge system, or a knowledge institution.

2.3 Data Science Defined

Biobanks do not exist solely with respect to the material samples collected. The information or data associated with each sample is itself a critical and related knowledge resource, and it should be considered for purposes of analysis jointly with the material to which it relates. The scale of the information resources collected in biobanks, particularly population-based biobanks for genetic research, suggests the importance of combining modes of analysis relative to specimen collection itself with modes of analysis as understood in the related field of data science, sometimes referred to as data-driven science or data-intensive science. For present purposes, references to data science refer to statistical and other analysis of data sets typically collected, stored, and processed in digital forms. Data models, the analytic frameworks used to define what data is collected and stored and the formats in which it is stored and shared, are as critical to data science as the data themselves. The popular phrase "Big Data" captures a part of what is distinctive about data science in modern biobanking and other scientific contexts, in the sense that the phrase speaks to the scale and speed at which information is collected in extra-large data sets and, in principle, made available to researchers. Data science extends beyond such large collections to include the systematic study of data in any

---

4 Henry Plotkin, *Darwin Machines and the Nature of Knowledge* (Cambridge, MA: Harvard University Press 1997).

form.[5] The point here is that biobanking governance is necessarily linked to data governance.

2.4 Governance Defined

The challenges of biobank classification illustrate the corresponding challenge of analyzing biobank governance. Governance is the object of analysis rather than (or, to be clear, in addition to) law as such. Governance refers to the multiple relationships among various institutional actors, from individuals to governments, and a specified resource or set of resources. Those relationships may consist of regulation, or discipline, or other modes of control or alignment. In most instances of interest here, governance addresses one or more social dilemmas associated with management of the resource, that is, with the fact that the resource, taken in isolation, does not "manage itself" as an autonomous "thing." Those dilemmas arise from the fact of potential conflicts between multiple objectives represented by different actors and institutions with interests in the resource. Knowledge sharing occurs on several dimensions simultaneously: temporal, generational, geographic, disciplinary, and beyond. Not every actor and not every objective can be fully accommodated at all times. When governance succeeds, it mediates and enables knowledge institutions, including biobanks, to thrive. Sometimes, governance does not succeed; sometimes, governance is designed and interposed intentionally but fails to accommodate competing objectives of different actors. Sometimes, governance emerges and evolves, more or less organically, and the processes of adaptation produce less and more thriving.

Like virtually all institutions for generating, storing, or otherwise managing knowledge resources, biobanks operate at the intersection of multiple, overlapping regulatory or disciplinary frameworks, which themselves operate at multiple levels. Formal positive law and regulation is only one source of institutional order. Even that formal law may come from multiple legal domains (intellectual property law, privacy law, and antitrust or competition law, for example) and from multiple institutions and sources (legislatures, courts, administrative or regulatory bodies). Informal frameworks (communities of practice, social norms, ideologies) may play important roles in disciplining both individual and collective behavior. Formal and informal systems each may be expressed via well-defined institutions, such as firms, universities,

---

[5] For example, Christine L. Borgman, *Big Data, Little Data, No Data: Scholarship in the Networked World* (Cambridge, MA: MIT Press 2015); Matthew J. Salganik, *Bit by Bit: Social Research in the Digital Age* (Princeton, NJ: Princeton University Press 2017).

and other legal entities; via markets; and via individuals who internalize relevant expectations. Both formal and informal disciplinary rules, norms, and expectations may be relevant with respect to scientific researchers who deposit, withdraw, and access and use biobank samples and information; with respect to information scientists who design and maintain the biobank itself; and with respect to computer scientists and programmers who maintain the information technology and network infrastructure that is typically associated with a biobank.

Law, policy, design, culture, economics, and ethics are combined, by necessity. What may be called "origin stories" of biobanks often play a central role in determining the narrative framework within which a given biobank exists, thus influencing the choice of relevant legal or disciplinary frameworks for application and analysis. My choice above to invoke the historical example of the small "physic garden" as an early example of a "biobank" itself suggests a historical narrative that distinguishes my summary, in part, from an alternative summary that might have begun with large modern gene banks.

Part of the concern for any governance analysis, in short, is scale. With respect to biobank resources, what sorts of collectives matter for purposes of preserving, accessing, and using the knowledge? Part of the concern is temporal. How should biobanks blend current interests in knowledge with future interests? Part of the concern is hierarchical or sequential. At times, it may be helpful to characterize biobanks as forms of knowledge infrastructure, in the sense that biobank resources offer a broadly distributed and widely shared knowledge resource as an input into a diverse array of potential knowledge outputs. When and how should biobanks support or enable other research? In what respects, if any, should biobanks be concerned with supporting or enabling not only other research but also applications of biobank resources for direct human benefit?

Resolutions of conflicts embedded in each of these concerns may not always align. In the specific instance, therefore, the research lens broadens: What helps a biobank thrive? What undermines a biobank? These are not purely legal or public policy questions; as this chapter has already emphasized, they are, in a broad sense, governance questions. The chapter next turns to exploring how those questions may be researched in a systematic and comparative way. What commonalities are relevant, and what case-specific circumstances matter?

3 THE KNOWLEDGE COMMONS RESEARCH FRAMEWORK

Connecting the conceptual and practical definitions of the preceding section—biobanks, knowledge, data science, and governance—yields the intermediate proposition that biobanks are knowledge commons institutions. The term "commons" should not mislead. As used here, commons refers to institutional arrangements for managing shared access to a pooled or collected resource. In adopting the knowledge commons characterization for biobanks, the chapter follows usage by other scholars.[6] The reason for identifying biobanks as knowledge commons is that doing so sets a foundation for the claim that understanding biobanks governance is a matter for empirical research. That research should be systematic and comparative across multiple biobank cases studies, to identify commonalities among the diversity of biobank forms.

This section describes one such approach to the relevant research: comparative institutional analysis using the knowledge commons research framework. The framework supplies an analytic basis for systematic comparative analysis of knowledge-sharing institutions of all kinds. The framework is set out in brief in this section. Biobanks are one leading illustration. The question for researchers is not "is a biobank a commons?"; rather, the question is, "are biobanks governed as commons?" What follows summarizes relevant research to date on biobanks as knowledge commons.

3.1 Background

Following Elinor Ostrom's groundbreaking work on institutions for resource management in the natural resource and environmental contexts,[7] Frischmann, Madison, and Strandburg describe knowledge commons generally as governance solutions for shared resources subject to social dilemmas.[8] In the first place, a resource is identified or created; next, use of that resource is purposefully shared by some population of producers and/or consumers. In the second place, a number of possible social dilemmas exist that are associated with the shared production and/or use of that resource, deriving generally from interests in social collaboration and cooperation. Commons address one or more of those dilemmas. Commons are forms of governance, or management, of shared resources.

---

6 Andrea Boggio, "Population Biobanks' Governance: A Case Study of Knowledge Commons," in Katherine J. Strandburg, Brett M. Frischmann, and Michael J. Madison (eds), *Governing Medical Knowledge Commons* (Cambridge: Cambridge University Press 2017); Barbara J. Evans, "Genomic Data Commons," in Katherine J. Strandburg, Brett M. Frischmann, and Michael J. Madison (eds), *Governing Medical Knowledge Commons* (Cambridge: Cambridge University Press 2017).

7 Elinor Ostrom, *Understanding Institutional Diversity* (Princeton, NJ: Princeton University Press 2005); Elinor Ostrom, *Governing the Commons: The Evolution of Institutions for Collective Action* (Cambridge: Cambridge University Press 1990).

8 Brett M. Frischmann, Michael J. Madison, and Katherine J. Strandburg (eds), *Governing Knowledge Commons* (Oxford: Oxford University Press 2014).

With respect to natural resources, Ostrom and her colleagues and collaborators demonstrated the viability of a range of sustainable, durable commons governance strategies that preserve the resource over time, implemented by local groups and communities using well-structured convention and custom. Commons are collectively managed governance systems, often marked by the absence of formal, market-based property law systems. In their repurposing of Ostrom's work, Frischmann, Madison, and Strandburg set out a research framework to investigate the viability of equivalent commons governance strategies with respect to knowledge, scientific, and cultural resources.[9] Knowledge commons governance may differ from natural resource commons governance in key respects, beginning with the fact that knowledge commons resources, unlike forests or fisheries, are naturally nonrival or nondepletable and therefore naturally or inherently shareable. The case for sustainable commons governance is neither inherently stronger nor weaker as a result. Instead, cases of knowledge commons must be researched from the beginning, rather than analyzed solely by analogy to natural resource commons. The point of the research framework presented here, like that of any research framework, is to permit research and data collection to proceed under a common set of assumptions and questions, even if specific research methods and disciplinary foundations may vary from researcher to researcher or from field to field. The framework is neither theory nor model. Strong theorizing and modeling may follow the research but only light and tentative theorizing, if any, should precede it.

The framework as described is borrowed from a 2014 book by Frischmann, Madison, and Strandburg titled *Governing Knowledge Commons*.[10] That book presents the framework and applies it to a set of case studies of institutions defined in part by knowledge-sharing practices with respect to one or more knowledge resources. While the framework is designed for application at the institutional level, referred to as commons, the intuitions and preliminary investigation that animated its development are applicable more broadly. Knowledge resources come in many forms; governance comes in many forms. Information and knowledge are principally immaterial, intangible resources, but they may be embodied in material forms, in flows of knowledge as well as in forms, and in labor and skill and time, as well as in embodied creation and other materials. The balance of this section gives a fuller account of the framework.

The knowledge commons framework builds on a series of related intuitions. Commons governance means knowledge and information management char-

---

9 *Ibid.*

10 *Ibid.*

acterized by domains of managed openness and sharing of relevant resources. The first intuition is that commons governance is in broad use in day-to-day practice in a variety of domains and across a variety of scales. Documenting evidence to justify that intuition is a primary goal of the framework.

The second intuition is that such structured openness in the management of both natural and cultural resources is likely to lead to socially beneficial and/or socially productive outcomes. Salient among the class of cases where commons governance is successful and sustainable are contexts where social interest in positive spillovers from bilateral market transactions is high. Commons may sustain the production of spillovers when the market otherwise may not. Describing the commons framework in terms of spillovers from bilateral market transactions runs a substantial risk of characterizing an information or knowledge context exclusively in "scarce resource" terms rather than in "abundant resource" terms. Care must be taken in applying the framework to understand the nature of the resources in question.

The final intuition is that a standard framework for identifying and assessing commons across a variety of domains can support the development of more sophisticated tools for realizing the potential for commons solutions in new institutional settings. It can also help to distinguish commons solutions from other solutions, such as an approach grounded solely in formal IP law, or in formal privacy law or contract law, which might be preferred. Applying the knowledge commons research framework is an exercise in analyzing colloquial commons institutions, such as "scientific research" taken in the aggregate, in a nuanced way, and in application to concrete examples.

Examining commons in knowledge and information contexts builds on the framework pioneered by Ostrom and her colleagues—known as the Institutional Analysis and Development (IAD) framework—but it adds some important modifications. The IAD framework has been used principally to structure analysis of solutions to collective action problems in natural resource contexts (so-called action arenas, or action situations) such as forests, fisheries, and irrigation systems.

First, the knowledge commons framework differs from the IAD framework in certain key respects. Unlike the IAD framework, it does not assume the agency of rational, choice-selecting, self-interested individuals, as the IAD framework tends to do. The social dilemmas subject to analysis in the knowledge commons context arise not only because of conflicts among rational, self-directed actors seeking to maximize their own benefits from resource access and use. Relevant social dilemmas arise because of conflicts among actors with diverse backgrounds, interests, and motivations. The knowledge commons framework accepts the role of historical contingency and of both inward- and outward-directed (selfless or other-oriented) agents in the evolution of collective or commons institutions.

Second, unlike natural resource commons, which largely take the existence of their resources for granted—fish, trees, water, and the like—knowledge commons identify resource design and creation as variables to be described and analyzed. As intellectual resources (that is, as forms of knowledge and information), patents, copyrights, and underlying inventions, creations, and data, and related material objects, are shaped by a variety of institutional forces rather than by nature.

Third, critically, the knowledge commons framework does not assume that the relevant resources are rival and depletable. The knowledge commons framework generally assumes precisely the contrary: that intangible information and knowledge resources are nonrival, nondepletable public goods. They may be closely linked, however, to depletable material resources. The dilemma to be solved is not primarily a classic "tragic commons" overconsumption problem. Instead, it is more likely (in part) an underproduction problem (how to produce the knowledge resource?) and (in part) a coordination problem. How may different actors coordinate their activities and interests in order to make appropriate and productive use of a shared resource?

The foregoing is not an exclusive list, however. In applying the framework to any particular case, care must be given to describing the authentic character of the social dilemmas present.

Finally, Ostrom's work largely ignored or discounted the role of the state in commons governance. The knowledge commons framework necessarily accepts the possibility that the state may have one or more key roles to play in managing shared knowledge resources.

3.2 The Framework

Against the background just set out, the knowledge commons framework proposes to undertake comparative institutional analysis by evaluating cases of commons resources via a series of questions, or clusters, to be applied in each instance.

Each case study investigation begins with a general description of the history and character of the problem that is being addressed by governance in the specific case or context. This may be an explanation that is internal to the governed institution(s) (problems and explanations may emerge from stories told by participants, either today or historically, or both), or an explanation that is external to the governed institution (such as the public goods account of the rise of IP law).

A researcher should ask whether the relevant resource or case is characterized from the outset by patent rights or other proprietary rights, as in the case of a patent pool, or by a legal regime of formal or informal openness, as in the case of public domain data or information collected in a government archive. That characterization influences the description of the social dilemmas addressed by the governance institution. A particular regime might involve securing the benefits of sharing data and information, or sharing rights in information, or sharing both. The character of the commons solution might involve encouraging the production of new resources, or coordinating holders of different property interests or holders of different public domain knowledge resources, for example.

Answering those initial questions sets a baseline against which a commons governance regime has been constructed. Within that regime, one next asks definitional questions. What are the relevant resources, taking into account both intangible and tangible resources and their individual or social character? What are the relationships among these resources, the baseline, and any relevant legal regime (for example, what a scientist considers to be an invention, what patent law considers to be an invention, and the boundaries of the patent itself are three related but distinct things)? What are the boundaries and constitution (membership) of the collective, community, or communities that manage access to and use of those resources? How is membership acquired (this may be informal, formal, or a blend of the two), and how is membership governed? What is good behavior within the group, and what is bad behavior? Who polices that boundary, and how?

Next are questions concerning explicit and implicit goals and objectives of commons governance, if any such goals and objectives exist. It is possible that commons governance regimes emerge from historical contingency rather than via planning. Is there an identified resource development or management dilemma that commons governance is intended to address, and what commons strategies are used to address that dilemma?

How "open" are the knowledge and information resources and the community of participants that create, use, and manage them? *Governing Knowledge Commons* argues that commons governance regimes involve significant measures of resource and community sharing and openness. The details of this openness should be specified in both absolute and relative terms, along with their contributions to the effectiveness of commons. Some commons and commons resources have precise and fixed definitions of both resources and community membership. Either resources or membership or both may be more fluid, with boundaries defined by flexible standards rather than by rules.

A large and critical cluster of questions concerns the dynamics of commons governance, or what Ostrom refers to as the "rules-in-use" of commons: the interactions of commons participants and resources. Included in this cluster of questions are: (1) details of stories of the origins, histories, and operations of commons; (2) formal and informal (norm-based) rules and practices regarding distribution and coordination of commons resources among participants, including rules for appropriation and replenishment of commons resources; (3)

the institutional setting(s), including the character of the regime possibly being "nested" in larger scale institutions and being dependent on other, adjacent institutions; (4) relevant legal regimes, including but not limited to property law; (5) the structure of interactions between commons resources and participants and institutions adjacent to and outside the regime; and (6) dispute resolution and other disciplinary mechanisms by which commons rules, norms, and participants are policed.

At this point the attributes of the system have been specified, and it becomes possible to identify and assess outcomes as the system operates in practice. In Ostrom's IAD framework, outcomes are typically assessed in terms of the resources themselves. Has a fishery been managed in a way that sustains fish stocks over time? Do commons participants, such as the members of a fishing community, earn returns in the commons context that match or exceed returns from participation in an alternative governance context, such as private market transactions? In knowledge commons settings, resource-based outcome measures may be difficult to identify and assess. Sustaining the products of the resource, individually or in combination, may be the point. A patent pool may serve as knowledge commons governance for a particular industry, but the success of the pool is not only measured by the fact that the pool itself survives; equally important, the value of the pool may be shown by the production of valuable complex products that could not be produced but for the pooling arrangement. Outcomes take different forms. It may be the case that social patterns of participant interaction constitute relevant outcomes as well as relevant inputs. What may matter is that the community itself thrives, in addition to the knowledge resource that it manages. Levels and types of interaction and combination matter. Participant interaction in the context of a shared resource pool or group may give rise to (or preserve, or modify) an industrial field or a technical discipline. In that specific case, such spillovers may be treated as relevant outcomes.

Having identified relevant outcomes, it becomes possible to look back at the dilemmas that defined commons governance in the first place. Has the regime solved those problems, and if not, then what gaps remain? How do the outcomes produced by commons governance differ from outcomes that might have been available if alternative governance had been employed? Has commons governance created costs or risks that should give policymakers and/or institution designers pause? Costs of administration might be needlessly high; costs of participation might be high. A collection of industrial firms that pool related patents in order to produce complex products may engage in anticompetitive, collusive behavior. Commons governance may facilitate innovation. It may also facilitate stagnation.

In sum, the knowledge commons framework provides a useful method of blending standardization and local adaptation in a systematic way in researching governance of knowledge institutions, such as biobanks, using comparative institutional analysis. From that foundation, the framework then provides the means to undertake more focused queries in order to define the relevant opportunity set for legal/regulatory analysis (that is, specifically with regard to law or policy approaches within the overall governance context). Finally, and most optimistically, results from using the framework may permit specifying a useful set of guidelines and recommendations for further development and design of knowledge institutions themselves. It is important to recognize the study of biobank governance as an opportunity to innovate regarding modes of governance beyond formal, public, or positive law; beyond publicly enacted regulation; and even beyond formalized "public/private partnerships," as the default or primary modes of encouraging and sustaining these institutions.

3.3 Applications: Biobanking Governance as Knowledge Commons

The claim presented in this chapter, that biobank governance should be analyzed as knowledge commons governance, is not novel. Since the initial publication of the knowledge commons research framework in 2010,[11] several researchers have adopted that framework in order to study biobanks and related institutions for collecting and managing biological specimens and associated data. This section reviews the relevant literature briefly, to show in part the viability of this approach and to show in part (by contrast) how other biobank governance literature might be adapted and brought within its scope. This is not an exhaustive literature review or bibliography; it advances the claim by illustrating it.

The most direct applications of the knowledge commons framework appear in case studies published in the edited collection by Strandburg, Frischmann, and Madison titled *Governing Medical Knowledge Commons*.[12] Four contributors to that volume have written case studies that address biobanks or features of biobanks using the knowledge commons governance framework. Peter Lee contributed "Centralization, Fragmentation, and Replication in the Genomic Data Commons," a treatment of methods and technologies used to manage information already contributed to large-scale genomic databases.[13] Barbara

---

11 Michael J. Madison, Brett M. Frischmann, and Katherine J. Strandburg, "Constructing Commons in the Cultural Environment" (2010) Cornell Law Review 95: 657.

12 Katherine J. Strandburg, Brett M. Frischmann, and Michael J. Madison (eds), *Governing Medical Knowledge Commons* (Cambridge: Cambridge University Press 2017).

13 Peter Lee, "Centralization, Fragmentation, and Replication in the Genomic Data Commons," in Katherine J. Strandburg, Brett M. Frischmann, and Michael J. Madison

Evans wrote a detailed case study of management of genomic databases composed of the results of consumer-driven genomic testing, "Genomic Data Commons."[14] Andrea Boggio offered "Population Biobanks' Governance: A Case Study of Knowledge Commons," whose title directly describes its content.[15] Jorge Contreras contributed "Leviathan in the Commons: Biomedical Data and the State," which directs attention to the key role that state support or coordination may play in commons governance contexts.[16] Jorge Contreras has also published a number of other studies of governance of genomic commons institutions, typically within the knowledge commons framework.[17]

Particular attention should be paid to the impressive recent work, *Governing Digitally Integrated Genetic Resources, Data, and Literature: Global Intellectual Property Strategies for a Redesigned Microbial Research Commons*, by Jerome Reichman, Paul Uhlir, and Tom Dedeurwaerdere.[18] The authors provide a comprehensive account of a central set of biobanking institutions as knowledge commons governance, addressed to the material and data-related results of microbial research. The work is especially valuable in the biobank context because its attention to plant genetic information complements the commons-based study of institutions to manage human biological material. A useful volume that anticipates many of the themes of biobank governance as knowledge commons is *Gene Patents and Collaborative Licensing Models: Patent Pools, Clearinghouses, Open Source Models and Liability*, edited by Geertrui van Overwalle.[19] That collection offers a number of useful perspectives on knowledge and information sharing with respect to biological data, though it does not consolidate them in a systematic way. A recent, important work on the challenges of Big Data and data science, with attention given

(eds), *Governing Medical Knowledge Commons* (Cambridge: Cambridge University Press 2017).

[14] Evans, *supra* note 6.

[15] Boggio, *supra* note 6.

[16] Jorge Contreras, "Leviathan in the Commons: Biomedical Data and the State," in Katherine J. Strandburg, Brett M. Frischmann, and Michael J. Madison (eds), *Governing Medical Knowledge Commons* (Cambridge: Cambridge University Press 2014).

[17] For example, Jorge Contreras, "Constructing the Genome Commons," in Brett M. Frischmann, Michael J. Madison, and Katherine J. Strandburg (eds), *Governing Knowledge Commons* (Oxford: Oxford University Press 2014).

[18] Jerome H. Reichman, Paul F. Uhlir, and Tom Dedeurwaerdere, *Governing Digitally Integrated Genetic Resources, Data, and Literature: Global Intellectual Property Strategies for a Redesigned Microbial Research Commons* (Cambridge: Cambridge University Press 2016).

[19] Geertrui van Overwalle (ed.), *Gene Patents and Collaborative Licensing Models: Patent Pools, Clearinghouses, Open Source Models and Liability Regimes* (Cambridge: Cambridge University Press 2009).

to biobanking and to the utility of Ostrom's work on commons governance in understanding data science for researchers in the twenty-first century, is *Big Data, Little Data, No Data: Scholarship in the Networked World*, by Christine Borgman.[20]

The foregoing research may be contrasted with research on biobanks and biobank governance that adopts specific frameworks for analysis rather than a wholesale comparative institutional framework. The work may be broken down generally into several categories, each of which has yielded important valuable contributions but none of which has produced a comprehensive, systematic vision of the domain. The point to emphasize is that none of these modes has yet been able to integrate multiple perspectives on the social dilemmas associated with biobanks in order to produce a complete and integrated legal and public policy analysis. The work neither describes the biobank landscape in full nor offers an overall guide to interpreting the strengths and weaknesses of specific existing biobanks and developing new ones.

One mode of research examines biobanks and biobank resources primarily in terms of questions of ownership, thus raising important questions about public accountability and access, proprietary right and incentives, and moral and ethical claims. Ownership interests and claims may be assessed both with respect to patients and consumers and with respect to scientific researchers and the owners or managers of the relevant biobank enterprise. Claims may be analyzed with respect to material specimens and also with respect to information derived from them. In specific legal terms, the questions include chattel property in the material specimens and patent, copyright, and data ownership with respect to the information resources. The vocabulary of commons may appear, though often in opposition to a term and concept borrowed from the law of property, "anticommons," which speaks to excessively fragmented property interests in a complex market context.[21]

A second mode of research examines biobanks from the standpoints of personal autonomy, privacy and security, and ethics, primarily with regard to patients, consumers, and other subjects of clinical trials, but also with respect to scientific researchers and even with respect to the ethical status of biological specimens and information. These are often not framed as property claims. Rather, in legal and public policy terms, questions may be posed in terms of

---

20 Christine Borgman, *Big Data, Little Data, No Data: Scholarship in the Networked World* (Cambridge, MA: MIT Press 2015).

21 Examples include the contributions in Giovanni Pascuzzi, Umberto Izzo, and Matteo Macilotti (eds), *Comparative Issues in the Governance of Research Biobanks: Property, Privacy, Intellectual Property, and the Role of Technology* (Springer 2013); Peter Lee, "Toward a Distributive Commons in Patent Law" (2009) *Wisconsin Law Review* 917.

transparency and consent. This literature is unlikely to highlight commons governance or knowledge sharing as a priority. Rather, the research question is typically how to address or accommodate individual and personal privacy interests within an institutional environment that is set up to pool and share information.[22]

A third mode of research focuses primarily on social norms and technical resources concerning information and knowledge production and sharing within the scientific community. The literature tends to situate proprietary claims and autonomy claims within the broader information collection and production environment, highlighting the normative value of scientific collaboration and the production of new scientific and medical knowledge. Commons governance may be discussed in this context, though often as a normative claim rather than as an analytic framework.[23]

A final mode of analysis of biobanking governance applies traditional styles of governance thinking, focusing on positive law within national legal systems, to biobanks and to biobank enterprises.[24]

The knowledge commons framework, developed through case studies over time, offers the prospect of integrating these results and analyses in a systematic way.

4 RESEARCH AND GOVERNANCE CHALLENGES: LESSONS TO DATE, AND THE FUTURE

This section sets out critical areas of future inquiry with respect to biobank research, both within the knowledge commons research framework and for possible use by policy analysts and institutional designers. To date, case studies of knowledge commons governance for biobanks are too few in

---

[22] Examples include Kris Dierickx and Pascal Borry (eds), *New Challenges for Biobanks: Ethics, Law and Governance* (Intersentia 2009); Henry T. Greely, "The Uneasy Ethical and Legal Underpinnings of Large-Scale Genomic Biobanks" (2007) Annual Review Genomics and Human Genetics 8: 343; Mark A. Rothstein and B.M. Knoppers, "Regulation of Biobanks" (2005) Journal of Law, Medicine and Ethics 33(1): 1; Mark Stranger and Jane Kaye, *Principles and Practice in Biobank Governance* (Routledge 2009).

[23] A recent example is Helen Yu, "Redefining Responsible Research and Innovation for the Advancement of Biobanking and Biomedical Research" (2016) Journal of Law and the Biosciences 3(3): 1.

[24] Examples include Herbert Gottweis and Alan Petersen (eds), *Biobanks: Governance in Comparative Perspective* (New York: Routledge 2008); Jane Kaye, Susan M.C. Gibbons, Catherine Heeney, Michael Parker, and Andrew Smart, *Governing Biobanks: Understanding the Interplay between Law and Practice* (Hart Publishing 2012).

number to conclude with certainty that any key legal reforms are necessary or that developers or managers of biobanks should follow any mandatory guidance. The research so far does suggest some key areas of focus in both respects.[25] These are set out in what follows.

First, it is perhaps most important to understand the various goals and purposes associated with a given knowledge institution, such as a biobank, as well as how those goals and purposes have evolved and how they relate to one another. Researchers may be tempted to make assumptions about those goals and purposes and to move directly into questions regarding property rights or privacy and ethics or the role of the state. That is almost certainly an error. The knowledge commons framework suggests that explicit goals need not be taken at face value; it is worth exploring implicit conflicts and dilemmas that are addressed by institutional structures.

Second, the relevant actors may be more numerous and their roles more complex than they first appear. In turn, that means that the relevant goals of the biobank and their interdependencies may also be complex. Interdependencies may yield additional goals and opportunities, or may yield barriers and limitations. Contributors, users, other researchers, managers, patients, subjects, host institutions, funders, external reviewers, and their respective institutional linkages—past, present, and anticipated—should be described. Formal, informal, and normative social and cultural structures should be mapped, changes over time should be detailed, and the different potential of planned and emergent behavior should be considered. Social hierarchies may matter more or less. Shared or distinctive cultural values may play important roles in organizational function. Intrinsic motivations for action may play important roles along with extrinsic ones. Detailed mappings may be necessary to ensure that the full portrait of the institution is described. For example, future access and use may be as important or more important than present access and use. Research use and clinical use may be more important than stewardship and preservation or heritage concerns, or may coexist, though with different values.

Third, the character of the resources may be complex as well. This chapter has highlighted the interdependencies of biobank resource pools that consist primarily of physical or material specimens and pools that consist primarily of intangible information or knowledge resources. Those conceptual interdependencies may correlate with interdependencies regarding technical systems to be developed to collect, store, manage, and access them. They may be constituted

---

[25] The following is adapted from Katherine J. Strandburg, Brett M. Frischmann, and Michael J. Madison, "Governing Knowledge Commons: An Appraisal," in Katherine J. Strandburg, Brett M. Frischmann, and Michael J. Madison (eds), *Governing Medical Knowledge Commons* (Cambridge: Cambridge University Press 2017).

as separate but related resource pools, or as integrated resource pools. They may be characterized with differing levels of item-specific identifiability and access/use parameters, and with interdependencies regarding governance strategies concerning access and use.

Many of those interdependencies arise from the multifaceted aspects of the problems of designing and managing a biobank. A key element of knowledge commons resources is their constructed character. That means that law, policy, and practice may play important roles in determining the form and identity of the resource itself and/or of resource units within it. Law, in this sense, may be definitional as well as regulatory in determining what "counts," in multiple senses, for purposes of inclusion, extraction, and use of resources within the biobank. That fact may or may not distinguish the character of a resource pool that consists primarily of material specimens from a resource pool that consists primarily of research data. "Things" as resources are designed by human engagement (law, policy, science), along with systems to govern them and those who use them.[26]

Fourth, the knowledge commons research framework, like Ostrom's work on natural resource commons, emphasizes close study of the interactions among relevant actors with respect to the resource, under conditions specified by formal and informal rules. These interactions take place in "action arenas," which may be physical, virtual, or conceptual. Any given knowledge institution may consist of and support multiple action arenas. Action arenas may be centralized or consolidated, and they may be distributed geographically or virtually. Understanding the action arenas of a given case may require research of an almost ethnographic character. At this point, the empirical character of the knowledge commons perspective comes into sharpest focus, because the behavior of various actors in actual practice cannot be assumed to correspond to conceptual frameworks associated with law or any other research discipline, such as economics. As we know from ordinary experience but as we often do not believe as researchers, people do not necessarily act in rational self-interest.

Fifth, assessing the success or failure or other standard of viability of a knowledge commons institution is critical, but methods and standards for doing so in the knowledge and information context are badly underdeveloped. This is a part of the field of commons research particularly, and of comparative institutional analysis in general, that needs the most effort.

---

[26] Michael J. Madison, "IP Things as Boundary Objects: The Case of the Copyright Work," (2017) LAWS 6(3): 13, doi:10.3390/laws6030013; Michael J. Madison, "Law as Design: Objects, Concepts, and Digital Things," (2005) Case Western Reserve Law Review 56: 381.

5 CONCLUSION

The primary points made by this chapter are as follows.

Whether at local, national, regional, or global level, biobanks are knowledge institutions. They collect, curate, and steward biological materials and associated knowledge and information for the benefit of future generations as well as for present scientific researchers. They house knowledge resources, and they provide important knowledge infrastructure for the production of new knowledge.

Biobank resources consist of more than the physical specimens they collect. Increasingly, those specimens are accompanied by critical information and data and/or are media that express critical information and data. Biobank governance is intertwined with data and information governance, and with data science. Law, policy, design, culture, economics, and ethics are combined, by necessity.

Biobank governance consists principally of various modes of knowledge and information sharing with respect to the resource that constitutes any particular biobank. Knowledge sharing occurs on multiple dimensions simultaneously: temporal, generational, geographic, disciplinary, and beyond. As a result, governance is required to enable biobanks to thrive amid multiple possible social dilemmas, meaning potential conflicts between multiple goals and interests represented in the resource.

Systematic study of knowledge-sharing institutions, such as biobanks, is most effectively conducted using a standard analytic framework for comparative institutional analysis that is tailored to the dynamics of knowledge. Here, the chapter proposes use of the knowledge commons research framework. Examples and illustrations are given of the framework as applied to biobanks. Preliminary research results are described, and recommendations for future research are suggested.

PART II

Biobanks, translational medicine, and tech transfer

3. Biobanks as innovation infrastructure for translational medicine

W. Nicholson Price II[1]

Biobanks represent an opportunity for the use of big data to drive translational medicine. Precision medicine demands data to shape treatments to individual patient characteristics; large datasets can also suggest new uses for old drugs or relationships between previously unlinked conditions. But these tasks can be stymied when data are siloed in different datasets, smaller biobanks, or completely proprietary private resources. This hampers not only analysis of the data themselves, but also efforts to translate data-based insights into actionable recommendations and to transfer the discovered technology into a commercialization pipeline. Cross-project technological innovation, development, and validation are all more difficult when data are divided between different biobanks and other data repositories.

One way to conceive of biobanks and the big medical datasets they create and embody uses the lens of infrastructure: how can biobanks and their data serve as infrastructure to support later innovation? Some efforts already fit into this model; for example, the United States' Precision Medicine Cohort—now renamed All of Us—aims to create a large, uniform dataset to be used for widespread future research. Other biobank-related data efforts, like Myriad's dataset on BRCA1/2 genetic variations, still function as entirely private resources. Treating medical big data as infrastructure has implications for how they should be governed, and suggests advantages to centralized control and relatively broad access. More broadly, viewing biobank-related data as infrastructure would place them at a distinctly earlier point in the commercialization pipeline, serving more to facilitate later steps in translational medicine rather than being viewed as potentially commercializable products themselves.

This chapter is divided into two parts. In the first, I briefly describe big data in medicine: the sources of medical data, the promises of medical big data,

1 For helpful comments and conversations, I wish to thank Ana Bracic, Rebecca Eisenberg, Brett Frischmann, and Timo Minssen. Rebecca Kaplan provided excellent research assistance. All errors are my own.

and a key challenge: data fragmentation. In the second, I discuss the role of biobanks in medical big data, focusing on their role in infrastructure for innovation and their potential for facilitating translational research.

1 BIG DATA IN MEDICINE

Big data has long been heralded as the next big revolution in health care—but that revolution has been relatively slow to arrive. Although data are constantly and increasingly generated from many sources of medical information, including research and samples associated with biobanks, those data are often fragmented into segments that are less useful than might be the whole. This section briefly describes the sources of medical data, the potential benefits of such data, and the challenge of fragmentation.

1.1 Sources of Medical Data

Big health data come in many forms. The most traditional, of course, are the health records generated in routine medical encounters, and now captured in electronic health records (EHRs).[2] These include doctors' notes, test results, patient medical history, diagnoses, and other medical information.[3] Insurance claims records, raw diagnostic testing data, and prescription records increase the picture of medical data. Less traditional, but increasingly a part of the picture, are the health-related data collected by wearable devices (medical or otherwise), including fitness trackers, insulin monitors, and smartphones.[4] Finally—and especially important in the context of this work—research data and patient samples, while only available for a subset of patients, provide extraordinarily deep data for that set. They often aim to provide an especially complete set of medical information for a particular patient because of the potential to answer questions that might arise later.

1.2 Promised Benefits

Big data promise substantial benefits for the health system. In the short term, they are supposed to help drive efficiency in health systems, and to show patterns of care, how practices can be improved, and the like.[5] But the bigger

---

[2] *See generally* SHARONA HOFFMAN, ELECTRONIC HEALTH RECORDS AND MEDICAL BIG DATA: LAW AND POLICY (1 edition ed. 2016).

[3] *Ibid.*

[4] *See* W. Nicholson Price II, *Regulating Black-Box Medicine*, 116 MICH. L. REV. 421 (2017).

[5] *Ibid.*

promise comes from future potential for innovation. Precision medicine promises to tailor care to individuals based on their individual characteristics. Some such relationships can be painstakingly and explicitly derived, leading to hypotheses testable through classical clinical trials. Other, more challenging methods rely on using truly vast sets of data and turning machine-learning algorithms loose on those datasets to find complex, implicit patterns.[6]

1.3 The Challenge of Data Fragmentation

Data fragmentation is a tremendous barrier to realizing the potential for medical big data, and the barrier on which this article focuses.[7] The promise of big data depends on linking data from multiple sources for an individual patient, and on linking data across many patients to determine useful patterns to direct innovation and care. Ideally, the available datasets would include comprehensive information for a broad set of patients. Unfortunately, data are generated by different sources and are often difficult to reunite. Primary care physicians, specialists, and others involved directly in care may maintain their own records, which are only sometimes linked. And even when data are linked across the spectrum of care, they are often unconnected from those data generated outside the context of care. Other data arise from research contexts, and may or may not be linked to clinical care data.[8] Biobanks may acquire both sources of data, as they can acquire both patient health records (or some fraction of such records) and data from research studies. Finally, some data arise from sources far from the health system, such as wearable devices or internet searches; these are currently unlikely to be linked with other health records save through the action of the patient in question. This fragmentation is exacerbated over time, as patients switch doctors, insurers, pharmacies, and wearable technologies, and join or drop out of research studies. Even a patient's primary

---

6 *See* W. Nicholson Price II, *Black-Box Medicine*, 28 HARV. J.L. & TECH. 419 (2015).

7 Other barriers certainly exist, and I do not mean to downplay them here. Data quality is a substantial hurdle. *See, e.g.*, Sharona Hoffman & Andy Podgurski, *Big Bad Data: Law, Public Health, and Biomedical Databases*, 41 J. LAW. MED. ETHICS 56 (2013). Other technological hurdles include storage and analyses of data. *See, e.g.*, Niels Peek et al, *Technical Challenges for Big Data in Biomedicine and Health: Data Sources, Infrastructure, and Analytics*, 9 Y. B. MED. INFORM. 42 (2014). For a description of other barriers in large scale observational research, *see, e.g.*, Rebecca S. Eisenberg & W. Nicholson Price II, *Promoting Healthcare Innovation on the Demand Side*, 4 J. L. & BIOSCIENCE 3, 23–39 (2017).

8 Sometimes legal barriers limit integration of research data into clinical care records, as when in the US a laboratory performs research but is not approved under the Clinical Laboratory Improvement Amendments (CLIA) to perform clinical testing.

care records can become time-fragmented if the patient is not diligent about having records transferred from one doctor to the next—and even if the patient is diligent, the lack of compatibility between different electronic health records may frustrate the merging of information.[9]

In addition to fragmentation of data within the records of an individual patient, of course, there is tremendous segregation of data from different patients. Doctors, hospitals, insurers, and others have little individual incentive to make their data available to those who would combine them into larger datasets—and in fact may be prohibited from doing so by privacy and security rules in many contexts.[10]

This fragmentation of data hinders the goals of big data in medicine, and limits the insights that can be derived.[11] Less comprehensive datasets limit the relationships that can be identified, and may lead to biased outcomes.[12] While overcoming fragmentation is not the only challenge to the use of big data to drive both basic and translational medicine, it is a substantial hurdle.

2 BIOBANKS, INNOVATION INFRASTRUCTURE, AND TRANSLATIONAL MEDICINE

Biobanks create possible avenues to use big data better in the context of translational medicine. Biobanks are well positioned to gather, generate, and store medical big data. By bridging the gap between basic research and real-world patient phenotypic samples and data, they can facilitate the translation of laboratory insights into clinical practice. In so doing, they play an infrastructure role, both *for* and *of* big data in health. By an infrastructure *for* data, I mean that biobanks can provide resources to store, transfer, analyze, and otherwise use data. But biobanks can also help create an infrastructure *of* data—that is, the data that biobanks create and store are themselves infrastructure for translational innovation. This section briefly addresses each of these issues.

---

9 *See, e.g.*, Andy Kessler, *Siri, Am I About to Have a Heart Attack*, WALL ST. J. (Jan. 9, 2017), www.wsj.com/articles/siri-am-i-about-to-have-a-heart-attack-1484007412 (noting challenges of EHR interoperability for medical big data and noting that Epic Systems, the leading provider of EHR systems, "appear[s] to be the leading obfuscator when it comes to transferring records and interoperability").

10 *See, e.g.*, Eisenberg & Price, *supra* note 6, at 34–9 (discussing the challenges to data integration posed by the Health Information Portability and Accountability Act (HIPAA)).

11 *See* W. Nicholson Price II, *Risk and Resilience in Health Data Infrastructure*, 16 COLORADO SCI. & TECH. L. REV. 65 (2017).

12 *See* Price, *Black-Box Medicine*, *supra* note 5, at 430–2 (2015) (describing the desirability of large datasets to identify complex relationships).

2.1 Biobanks and Medical Big Data

Biobanks occupy a special role in the universe of medical big data for at least three reasons. First, at least some biobanks collect data from at least two spheres of data—health care data and research data—that are often held separately. Biobanks are in the business of collecting both samples and data from patients;[13] the first are substantially less useful without the second. Thus, biobanks collect patient medical information along with samples.[14] To the extent that biobanks acquire research results about those samples, they have the advantage of aggregating both medical and research data about an individual. Researchers who actually analyze the samples also acquire both types of information, of course, but only for the patients in their own studies, while biobanks can potentially join information about many more patients represented in their collections. Meanwhile, in the process of acquiring and processing samples, biobanks may perform many analyses outside whatever research protocol was specified for gathering the sample in the first place; for instance, biobanks may sequence genomes, quantify mRNA populations, measure metabolite and/or protein levels, and histologically classify samples.

Second, because biobanks maintain collections of biological specimens, there exists the potential for performing currently unplanned analyses.[15] Uniquely among repositories of patient information, biobanks have the capacity to generate significant amounts of new data without acquiring it from individuals, by reanalyzing samples using new technology. To take an obvious example, consider a collection of tumor samples gathered throughout the course of several decades. For most of that time, the samples would not have been genetically analyzed because the technology was not available. But now, the entire set of samples could be genetically sequenced and the resulting sequence data could be linked to tumor pathology and other medical information about the patients that the biobank recorded. Sometimes this approach can create controversy, as with genetic analyses of blood spots collected from newborn infants. Ideally, such analyses are facilitated both by broad upfront

---

13 Indeed, some have suggested that merely by standardizing the collection of data and patients within the catchment of a biobank, patient care may already be improved. Conor M.W. Douglas & Philip Scheltens, *Rethinking Biobanking and Translational Medicine in the Netherlands: How the Research Process Stands to Matter for Patient Care*, 23 Eur. J. Hum. Genet. 736 (2015).

14 *See, e.g.*, Timo Minssen & Jens Schovsbo, *Legal Aspects of Biobanking as Key Issues for Personalized Medicine and Translational Exploitation*, 11 Pers. Med. 497 (2014).

15 *See, e.g.*, Gerardo Botti et al, *Tumor Biobanks in Translational Medicine*, 10 J. Transl. Med. 204, 204 (2012).

consent (or other models that permit ongoing consent) and by the addition of more recent health data from the individual, where available.

Third, biobanks—or at least some fraction of them—already have as part of their mission a role as the repository for information, whether embedded in biological specimens or found in biological data. They are created with the idea that they will collect samples and make those samples available to future researchers, along with associated data. Thus, biobanks already provide infrastructure for biomedical innovation. This role is acknowledged explicitly in some cases. For instance, the Austrian-headquartered Biobanking and BioMolecular resources Research Infrastructure (BBMRI)[16] is a "distributed research infrastructure of biobanks and biomolecular resources," which aims to connect researchers and biobanks and "facilitate the use of samples/data collected in Europe for the benefit of human health."[17] This raises the question: how exactly can biobanks help enable the use of samples and data for human health? A longtime answer is that biobanks can provide resources that are useful for basic research. But biobanks can also facilitate research later in the pathway.

2.2 Biobanks and Translational Medicine

Biobanks are a key resource for developing translational medicine. They help make the jump from basic research discoveries to the phenotypic reality of patient populations represented by samples and data.[18] In a meaningful sense, this is because biobanks themselves straddle the divide between basic and clinical research; they are established as tools to help initial research, but do so by collecting large amounts of real-world samples and data.

Biobanks can facilitate translational medicine in several ways. For instance, a basic lab discovery might identify a gene with potentially significant clinical impact because it encodes a protein that might be a potential drug target. Biobank samples and data can thus help demonstrate whether relevant gene variants are present in patients in the represented population, and can demonstrate real-world correlations with the disease of interest.[19] Biobanks can similarly be used to identify biomarkers to be used in drug development, and later to validate and quantify those same biomarkers.[20] Such biomarkers can

---

[16] BBMRI, *Frequently Asked Questions*, www.bbmri-eric.eu/faq/.

[17] *Ibid.*

[18] Minssen & Schovsbo, *supra* note 13.

[19] *See* Botti et al, *supra* note 14.

[20] *See* Arndt A. Schmitz, *Potential of Biobanking in Translational Medicine*, Presentation, HandsOn: Biobanks (Helsinki, Finland, 2014), *available at* http://

be prognostic (predicting the natural course of an illness),[21] predictive (helping identify a specific treatment), or pharmacodynamic (suggesting an optimal dose).[22] Finally, biobank data from patients who have already participated in clinical trials may be able to help stratify patients in those trials retroactively, and to develop new information about the already tested therapeutic agents.[23]

2.3 Biobanks as Data Infrastructure

Biobanks can serve an important role in providing data infrastructure for translational innovation in medicine. When I say data infrastructure, I mean both infrastructure *for* data—that is, resources for storing, collecting, and using data—and infrastructure *of* data—that is, the data themselves as infrastructure for later innovation.

Before getting into these two types of data infrastructure, it is worth being more explicit about what I mean by infrastructure. I principally adopt Brett Frischmann's characterization of infrastructure: (1) resources that can be "consumed nonrivalrously for some appreciable range of demand," which demand (2) is "driven primarily by downstream productive activities that require the resource as an input," where (3) those downstream productive activities result in a "wide range of goods and services, which may include private goods, public goods, and social goods."[24] The fact that infrastructure is widely usable for a broad range of outputs, some of which are public goods, implies that infrastructure is likely to be underprovided by private sources,[25] and also suggests that infrastructure resources are best kept relatively general to allow many uses rather than being specialized for one particular use.[26]

Biobanks are a promising source of infrastructure *for* data. By that I mean that they are designed to be repositories of samples for use by researchers, whether those researchers are at the early stages of discovery or later, in the process of translating fundamental insights into useful treatments that can be implemented in the clinic. Biobanks can serve a similar role—though

---

handsonbiobanks.org/documents/114074/129625/Schmitz_Biobanking_in_translational_medicine_HOBB2014.pdf/0dd2c1e9-298d-4af2-bc63-a34895442d7e.

[21] *See, e.g.*, Tobias M. Gorges & Klaus Pantel, *Circulating Tumor Cells as Therapy-Related Biomarkers in Cancer Patients*, 62 CANCER IMMUNOL. IMMUNOTHER. 931 (2013).

[22] *See* Schmitz, *supra* note 19. Biobanks can also supply samples not only to identify new biomarkers, but also to develop assays to measure those biomarkers. *Ibid.*

[23] *Ibid.*

[24] BRETT FRISCHMANN, INFRASTRUCTURE: THE SOCIAL VALUE OF SHARED RESOURCES 61–2 (2012).

[25] *Ibid* at 15.

[26] *Ibid* at 65.

even more explicit—for data related to such samples, or even more broadly. They provide the physical resources—freezers and collection equipment for samples, computers and networks for data—that let these valuable resources be collected, stored, accessed, and used. They can also provide intangible infrastructural resources, such as protocols for sample and data collection, patient procedures, or even norms about collection and use.[27] These resources may be tied directly to the biobank, or to the umbrella organization as a parallel to the sample-driven physical biobank.[28] One such example may be found in New Haven, where Yale University hosts the world's largest genomic biobank.[29] The biobank consists of specimens and data from more than 500,000 participants in the ongoing Million Veteran Program.[30] The biobank itself will store and maintain the data, providing an infrastructure for those data.

But biobanks are also important in the creation of an infrastructure *of* data—by which I mean that they generate, maintain, and make accessible information which is itself infrastructure that provides resources for future innovation.[31] This goal may be explicit; the Yale biobank, for instance, is best "viewed as a long-term infrastructure project" providing support for current and future researchers, according to its codirector.[32] The Precision Medicine Cohort (now All of Us), formed as part of President Obama's Precision Medicine Initiative, similarly aims to develop a very large dataset that can be used to support future innovation.[33]

How might such a broad infrastructure project work best? And what does the conception of infrastructure for biobanks gain us? Ideally, infrastructure

---

[27] *See, e.g.*, Nat'l Comm. on Vital & Health Stats, Information for Health: The Strategy for Building the National Health Information Infrastructure 11 (2001) (defining infrastructure for health data very broadly).

[28] Although I argue the mission and funding of data infrastructure and sample storage are similar, I recognize that some resources and forms of expertise differ between the two functions.

[29] John D. Curtis, *Million Veterans Program Now World's Largest Genomic Biobank*, Yale School of Medicine News (Aug. 8, 2016), https://medicine.yale.edu/news/article.aspx?id=13225.

[30] *Ibid.*

[31] *See* OECD, Data-Driven Innovation: Big Data for Growth & Well-Being, 177–206 (2015) (applying an infrastructure model for big data generally); W. Nicholson Price II, *Big Data, Patents, and the Future of Medicine*, 37 Cardozo L. Rev. 1401, 1439–44 (2016) (describing an infrastructure model for medical big data to support the development of complex medical algorithms to direct treatment).

[32] Curtis, *supra* note 28.

[33] Francis S. Collins & Harold Varmus, *A New Initiative on Precision Medicine*, 372 N. Engl. J. Med. 793 (2015) (describing the precision medicine initiative); Eisenberg & Price, *supra* note 6, at 44 (describing the Precision Medicine Cohort as government-provided innovation infrastructure).

for and of data in the biobank context should be connected, interoperable, and accessible.[34] These ideals arise because of the nature of infrastructure in enabling a broad range of different users and uses,[35] and are closely related to the FAIR Guiding Principles for data management laid out in 2016: findability, accessibility, interoperability, and reusability.[36] In playing an infrastructural role, not only should biobanks ensure that data are findable and reusable, but they should also proactively link data to make them usable in many contexts and future studies.

Connection means that individual players—in this case, individual biobanks—should be connected to each other, sharing resources and data.[37] This helps make the available datasets bigger and more comprehensive, which in turn enables the study of more complex relationships or rare conditions.[38] In addition, connection helps ensure that biobanks as a group facilitate broad and varied uses rather than focusing on uses specifically tailored to a particular use or user.

Interoperability is a key enabler of connection. That is to say, if biobanks store their data in different, mutually incompatible formats, connection becomes much more challenging.[39] Such interoperability challenges are already a major concern in the context of electronic health records.[40] To the extent that biobanks create their own data structures, interoperability concerns can swamp the possibility of meaningfully connected data. Policy efforts should therefore encourage the use of compatible data formats to better enable

---

34 For a broader description of several principles for data as infrastructure, *see* OECD, *supra* note 30, at 188 ff.

35 *See* Frischmann, *supra* note 23, at 61–2.

36 Mark D. Wilkinson et al, *The FAIR Guiding Principles for Scientific Data Management and Stewardship*, 3 Scientific Data 160018 (2016).

37 *See, e.g.*, Botti et al, *supra* note 14 (noting cooperation between biobanks allowing the study of rare cancers); OECD, *supra* note 30 (describing the need for connection).

38 Examples of more comprehensive datasets—whether through centralized or distributed architecture—are becoming more common. The FDA's safety surveillance Sentinel system, for instance, relies on a distributed architecture where data are kept by their creators but are available for centralized querying. *See* Susan Forrow et al, *The Organizational Structure and Governing Principles of the Food and Drug Administration's Mini-Sentinel Pilot Program*, 21 Pharmacoepidemiol. Drug Saf. 12 (2012); Eisenberg & Price, *supra* note 6, at 41–4 (describing the use of possible use of Sentinel or Sentinel-like systems to promote healthcare innovation by payers).

39 *See* OECD, *supra* note 30, at 192–94.

40 *See* Eisenberg & Price, *supra* note 6, at 25–6 (describing interoperability challenges).

a data infrastructure.[41] For instance, the Minimum Information About Biobank data Sharing model (MIABIS), developed by the BBMRI, aims to create a new "bio-object infrastructure" within the EU.[42]

Finally, to provide infrastructure for and of data, those data must be accessible to researchers. Finding the correct model of accessibility is not easy.[43] The costs of access may be substantial, and funding such access either by surcharges on researchers using the information or by other public/private mechanisms each have their own challenges.[44] Funding based on fees to users is the most straightforward possibility, but risks privileging larger market incumbents over new entrants, and undermines the infrastructure model of biobanks.[45] This is true because of the varied nature of output goods from infrastructure goods; users that create downstream public goods are unwilling to pay for access to the resource at a socially desirable level.[46] Similarly, those developing infrastructural goods are unlikely to invest at a socially optimal level because they cannot capture the full benefits.[47] Myriad Genetics provides a useful example of this dynamic in action: Myriad keeps its vast trove of health and genetic information on women who have used its BRCAnalysis service to test for mutations in the BRCA1 and BRCA2 genes, linked to breast

---

41 *See, e.g.*, U.S. Office of the National Coordinator for Health Information Technology, Connecting Health and Care for the Nation: A Shared Nationwide Interoperability Roadmap (Draft) 10–11 (2015), www.healthit.gov/sites/default/files/nationwide-interoperability-roadmap-draft-version-1.0.pdf.

42 Loreana Norlin et al, *A Minimum Data Set for Sharing Biobank Samples, Information, and Data: MIABIS*, 10 Biopreservation Biobanking 343 (2012). The development of the Data Sharing model has been a complex process. *See* Sakari Tamminen, *Bio-Objectifying European Bodies: Standardisation of Biobanks in the Biobanking and Biomolecular Resources Research Infrastructure*, 11 Life Sci. Soc. Policy 13 (2015).

43 *See* Kathleen Liddell & Johnathon Liddicoat, *Open Innovation with Large Bioresources: Goals, Challenges, & Proposals*, University of Cambridge Faculty of Law Research Paper No 6/2017 (2017), *available at* https://papers.ssrn.com/sol3/papers2.cfm?abstract_id=2888871.

44 *See* OECD, *supra* note 30, at 191–92.

45 *See ibid* at 15, 18–19 (discussing different models of covering costs); *cf.* Barbara J. Evans, *Sustainable Access to Data for Postmarketing Medical Product Safety Surveillance under the Amended HIPAA Privacy Rule*, 24 Health Matrix 11 (2014) (noting the challenges to funding for access to data under Health Insurance Portability and Accountability Act requirements forbidding adequately charging for access, and suggesting the availability of higher fees).

46 Frischmann, *supra* note 23, at 68–9.

47 *Ibid.*

cancer.[48] While the company invested enough to create a valuable resource of genetic data and sequences for its own use, its exclusionary business model meant that other users could not access those samples or data for socially beneficial purposes such as better understanding or confirmatory testing.[49]

Funding from other private or public resources, such as general tax revenues, is thus likely a better solution in terms of enabling broad access.[50] Infrastructure goods typically create substantial spillovers—indeed, that is much of their purpose—bolstering the case for public funding.[51] But public funding raises political economy concerns of procuring the funding in the first place and of—arguably—leaving money on the table once innovations have been developed.[52] This latter concern is the obverse of the spillover benefit: spillovers are, by definition, uncaptured benefits, and while those are a classic benefit of public spending, they can also create friction and concerns about properly managing the public risk. Some approaches might try to blend the public and private funding model to resolve this tension, perhaps requiring different access fees for different types of data users.[53] Setting the appropriate balance in such a blended approach may bring its own difficulties.

Funding is not the only challenge; balancing data accessibility against privacy is nontrivial.[54] As described above, broader access is important for an infrastructural good to enable various downstream uses. Nevertheless, individuals regard health data as sensitive, and thus privacy concerns arise, not only from the perspective of participant buy-in but also to satisfy legal and policy requirements. Various access models have been proposed to lower privacy risks, including models drawing from the literature on sharing data

---

48 *See, e.g.*, Robert Cook-Deegan et al, *The Dangers of Diagnostic Monopolies*, 458 NATURE 405 (2009).

49 *Ibid.*

50 Frischmann, *supra* note 23, at 94.

51 *Ibid* at 14.

52 *See* Rainer Warth & Aurel Perren, *Construction of a Business Model to Assure Financial Sustainability of Biobanks*, 12 BIOPRESERVATION BIOBANKING 389 (2014); Liddell & Liddicoat, *supra* note 42, at 18–19; *see* OECD, *supra* note 30, at 191–2.

53 Liddell & Liddicoat, *supra* note 42, at 12–13 (noting that the 100,000 Genomes Project in the UK applies differential licensing and IP terms to different entities seeking to use the Project's data).

54 *See* Roger A. Ford & W. Nicholson Price II, *Privacy and Accountability in Black-Box Medicine*, 22 MICH. TELECOMM. & TECH. L. REV. 1 (2016) (describing the tradeoff between privacy of patient data and the ability to verify the quality of complex medical algorithms developed using those data); INST. OF MED., SHARING CLINICAL TRIAL DATA: MAXIMIZING BENEFITS, MINIMIZING RISK (2015), *available at* http://nap.edu/18998 (extensively analyzing the sharing of clinical data and suggesting different models)

from clinical trials.[55] Technical mechanisms may also help to allow data access without reducing privacy, though these come with their own hurdles to adoption.[56] Despite the challenges, access is key; without access, the data risk staying in isolated silos that help promote neither early stage innovation nor later translational work.

In the present, these goals of connection, interoperability, and access are far from a reality, at least in many instances; biobanks and the data they house are highly fragmented, limiting their value as infrastructure resources.[57] There are at least several hundred biobanks across the globe, and perhaps thousands, depending on methodology and definitions.[58] Different biobanks are and were founded for different purposes, and to serve different patients—who may have had different views about acceptable purposes for their samples' retention, and who may have consented to different types of future research. Nonprofit and for-profit biobanks may have different motives, but still often keep resources fragmented. For-profit biobanks are driven by competitive forces to keep their samples and data tightly siloed and unavailable to others. The contours of intellectual property rights, trade secrecy, and different regulatory rules for different types of data may unfortunately encourage data silos of that

---

55 *See, e.g.*, Ford & Price, *supra* note 53, at 29–43.

56 *Ibid.*

57 Other problems arise from legal questions, in particular the issue of intellectual property rights and the desire for proprietary data. *See, e.g.*, Minssen & Schovsbo, *supra* note 13; Michiel Verlinden, Timo Minssen, & Isabelle Huys, *IPRs in Biobanking: Risks and Opportunities for Translational Research*, 2 Int. Prop. Quarterly 106 (2015). Those concerns will not be addressed here, other than to note that the desire to keep data and samples proprietary—or to exert strong intellectual property rights to limit future use or demand substantial compensation—cuts against the idea of biobanks as providing broadly accessible infrastructure for future innovation. *See* Liddell & Liddicoat, *supra* note 42. Toll roads may be useful to maintain a reasonable infrastructure, but excessive tolls slow the flow of useful traffic.

58 *See* Gregory J. Boyer & Warren Whipple, *Biobanks in the United States: How to Identify an Undefined and Rapidly Evolving Population*, 10 Biopreserv Biobank 511 (December 2012) (finding hundreds of biobanks and acknowledging the uncertainty of the count); R. Jean Cadigan, Dragana Lassiter, et al, *Neglected Ethical Issues in Biobank Management: Results from a U.S. Study*, 9 Life Sci. Soc'y Pol'y 1 (December 2013) (estimating about 800 biobanks in the United States); Hana Odeh et al, *The Biobank Economic Modeling Tool (BEMT): Online Financial Planning to Facilitate Biobank Sustainability*, 13 Biopreserv. Biobank. 421 (2015) (estimating thousands of biobanks); Bryan Keogh, *European Biobanks Forge Cross-Border Ties*, 103 J. Nat. Cancer Inst. 1429 (2011) (estimating tens of thousands of biobanks globally).

type.[59] Myriad again provides the poster child for such private data siloing.[60] Nonprofit biobanks are still driven by incentives that may encourage fragmentation, including access to grant funding, prestige, or focus on particular diseases, or national mandates. The biobank operated by Partners Healthcare in Cambridge, Massachusetts, for instance, provides samples only to about 6,000 researchers affiliated with Partners, for approximately $20 each, potentially raising concerns about access both in terms of cost and in terms of who can reach the resources.[61] In addition to the problem of mixed incentives for connecting and sharing data, biobanks that do wish to share data face other obstacles, including data regulations such as the HIPAA Privacy Rule in the United States and the General Data Protection Regulation in the European Union; the latter will impact both intra-EU data use and sharing of data between US and EU biobanks.[62]

Some biobanks are already addressing fragmentation concerns, such as the set of European biobanks that have adopted the MIABIS framework to help facilitate connection.[63] And other scholars are convening and working to address questions of fragmentation, connection, interoperability, and access.[64] But substantial challenges remain as many biobanks globally keep their resources proprietary and thus leave potential innovations in translational medicine, or other areas, on the table.

3 CONCLUSION

Biobanks hold tremendous possibilities to serve as innovation infrastructure for translational medicine. They generate and store both biomedical samples and data, and these resources can be used to help bring innovation from the basic research laboratory into clinical practice. Among other challenges,

---

59 *See, e.g.*, Arti K. Rai, *Risk Regulation and Innovation: The Case of Rights-Encumbered Biomedical Data Silos*, 92 Notre Dame L. Rev. 1641 (2017).

60 *See, e.g.*, Robert Cook-Deegan et al, *The Dangers of Diagnostic Monopolies*, 458 Nature 405 (2009); *but see* Dan L. Burk, *Patents as Data Aggregators in Personalized Medicine,* 21 B.U. J. Sci. & Tech. L. 233 (2015) (describing how Myriad used its patents to aggregate—and partially defragment—large amounts of data about breast cancer).

61 Beth Daley & Ellen Cranley, *The Rise of Bio-Rights: Patients Demand Cash for DNA Samples*, The Eye (October 10, 2016), https://eye.necir.org/2016/10/10/rise-bio-rights-patients-demand-control-get-cash-dna-samples/.

62 See Eisenberg & Price, *supra* note 6.

63 Roxana Merino-Martinez et al, *Toward Global Biobank Integration by Implementation of the Minimum Information about Biobank Data Sharing (MIABIS 2.0 Core)*, 14 Biopreservation Biobanking 298 (2016).

64 *See, e.g.*, Liddell & Liddicoat, *supra* note 42.

fragmentation of data reduces the ability of biobanks to play this role; in many instances, data are proprietarily maintained in biobank-specific silos. Biobanks should consider not only that they provide infrastructure for their data, but also that those data themselves can serve as infrastructure for forward-looking innovation. With better connection, interoperable data, and technical standards, and broad accessibility to researchers, biobanks and networks of biobanks can play a larger role in facilitating translational medicine.

4. Responsible use of human biosamples in the bioscience industries

Brian J. Clark and Tina Bossow[1]

1 INTRODUCTION

In the context of this chapter, "bioscience industries" refers to companies involved in the research, development, and manufacturing of products that improve diagnosis, treatment, or prevention of human diseases, or the provision of enabling technologies or services in support of the same aims.[2] This diverse grouping includes the pharmaceutical, diagnostics, biotechnology, and research and development services industries.

Many companies within the bioscience industries use biological samples derived from humans ("human biosamples"[3]) in their research, development, or manufacturing processes and some finished products may contain human biosample components. While the for-profit business model of commercial companies sometimes leads to distrust and questions over their legitimacy as

---

[1] Both authors are employees of Novo Nordisk A/S, a bioscience industry company that uses HBS. All views, analysis, and perspectives expressed in this chapter reflect the personal views and experiences of the authors, obtained in the course of current employment, previous employments, and other personal activities. They do not represent the views or opinions of Novo Nordisk A/S or any other organization.

[2] UK Bioindustry Association "What is biotech" www.bioindustry.org/about-us/what-is-biotech.html accessed October 12, 2017. In other contexts, the term "bioscience industries" also encompasses other types of biological industries involved in, for example, agricultural, food, microbiological, or biochemical research, development, and manufacturing. These are excluded from the context of this chapter. *See* Biotechnology Innovation Organisation, "About BIO" www.bio.org/about accessed October 12, 2017; *see* also BioOhio "What is bioscience" www.bioohio.com/ohio/whats-bio/ accessed October 12, 2017.

[3] Also known by other terms, such as human biospecimen, human sample, human specimen, human tissue, and others.

users of human biosamples (HBS),[4] such companies are and will remain essential for the discovery, development, and delivery of innovations in healthcare that are of benefit for populations. Therefore, the legitimacy of the bioscience industries as valid users of HBS for purposes intended to deliver public benefit should not be in doubt.[5] Rather, the pivotal question should be whether the bioscience industries can demonstrate that they use HBS in ways that are socially responsible, ethical, compliant with applicable laws or regulations, and safe, such that they merit access to precious HBS. In fact, this same question should equally apply to any user of HBS, whether a for-profit industrial user or a non-profit or public institutional user. All users should be able to demonstrate that they are legitimate and *responsible* users of HBS.

2 WHAT DO BIOSCIENCE INDUSTRIES USE HUMAN BIOSAMPLES FOR?

A variety of *ex vivo* uses of HBS exist. Human biosamples allow the user to recapitulate or mimic the biology of human health and disease states in the laboratory or manufacturing plant in ways that cannot practicably be achieved *in vivo* using human volunteers. They allow for uses at the organ, tissue, cellular, or molecular levels. Perhaps most importantly, they generate information or products that are directly human-relevant and do not rely on extrapolation from other models, such as experimental animals, where the degree to which the biology reproduces or predicts that of humans is questionable.[6]

The use of HBS can facilitate the search for and validation of new drug targets.[7] New drugs may be developed and tested in the laboratory for their stability, effectiveness, safety, routes of administration, metabolism and

4 Dianne Nicol and Christine Critchley, "What benefit sharing arrangements do people want from biobanks? A survey of public opinion in Australia" in Jane Kaye and Mark Stranger (eds), *Principles and Practice in Biobank Governance* (Routledge 2016).

5 *See*, for example, UK Biobank, "Principles of access" (UK Biobank, 2012) www.ukbiobank.ac.uk/principles-of-access/ accessed October 12, 2017.

6 Junhee Seok et al, "Genomic responses in mouse models poorly mimic human inflammatory diseases" (2013) PNAS 110: 3507–12; Adam C Drake, "Of mice and men: what rodent models don't tell us" (2013) Cellular & Molecular Immunology 10: 284–5; Robert L Perlman, "Mouse models of human disease: an evolutionary perspective" (2016) Evolution, Medicine, and Public Health 1: 170–6.

7 Kurt Zatloukal & Pierre Hainaut, "Human tissue biobanks as instruments for drug discovery and development: impact on personalized medicine" (2010) Biomark Med 4: 895–903.

clearance from the body.[8] In some cases highly purified HBS may become the therapeutic products themselves, for example, highly purified proteins or cellular/tissue therapies.[9]

New biomarkers of health and disease, which may be used for diagnosis, prognosis, or precise selection of treatments,[10] may be found in HBS.[11] New *in vitro* diagnostic tests may be developed and validated, some of which may contain HBS as critical ingredients.

As science and engineering progress, the potential for innovative and more powerful uses of HBS,[12] perhaps in ways previously unimaginable, is likely to increase. Therefore, vigilance and appropriate governance will be necessary for any organization to ensure that these new ways of using HBS are undertaken responsibly.

3 WHERE DO BIOSCIENCE INDUSTRIES OBTAIN HUMAN BIOSAMPLES?

Each company in the bioscience industries has its own particular sources and supply chains to obtain the HBS necessary for its work. However, some generalizations can be made.

It is less frequent that HBS are donated directly from individuals ("donors") to a bioscience company. However, when this does occur, it provides the company with the greatest degree of control over both the donation and all steps up to the use of HBS, hence strengthening the governance of the processes. In such cases, the donations often occur in the same jurisdiction as the use, so the advantage is that there is only one set of laws, regulations, and societal expectations to observe.

More typically, the HBS used by bioscience companies have to be obtained through one or more intermediaries between the donors and the companies who use the HBS. Examples of intermediary organizations are given in Table

---

8 Anthony Holmes, Frank Bonner, & David Jones, "Assessing drug safety in human tissues—what are the barriers?" (2015) Nat Rev Drug Discov 14: 585–7.

9 Proteins: for example, *Thrombate III*, a purified human Antithrombin III from pooled plasma samples; *Octagam*, a pooled immunoglobulin replacement therapy; *Abbokinase*, a Urokinase preparation derived from human neonatal kidney cells. Cellular/tissue therapies: for example, umbilical cord-derived haemopoietic stem cells.

10 Robert E. Hewitt, "Biobanking: the foundation of personalized medicine" (2011) Curr Opin Oncol 23: 112–19.

11 Christopher Womack & S. Rachel Mager, "Human biological sample biobanking to support tissue biomarkers in pharmaceutical research and development" (2014) Methods 70: 3–11.

12 Donald E. Ingber, "Reverse Engineering Human Pathophysiology with Organs-on-Chips" (2016) Cell 164: 1105–9.

Table 4.1 Examples of HBS supply intermediaries

Type of organization	Functions in the supply chain
Clinic, hospital, healthcare institute, academic health centre	Donor recruitment, HBS collection, clinical data annotation, HBS processing, storage, distribution, research collaboration/services
Biobank (commercial vendor, nonprofit/public/academic biobank)	Donor recruitment, HBS collection, processing, quality control, analysis, storage, distribution, research collaboration/services
Clinical or laboratory contract research organization	Donor recruitment, HBS collection, processing, quality control, analysis, storage, distribution, research services
Commercial distributor	Storage, distribution

4.1. The most typical intermediary is a healthcare provider, who interacts with the donors, receives the donated HBS, and provides the HBS onwards for use.

Human biosamples may also be obtained via "biobanks"[13]—organizations specifically managed to conduct any combination of the activities of donation, processing, storage, characterization, and onward supply of HBS and associated data,[14] most usually intended for use in research (Figure 4.1). Biobanks may be public, nonprofit, or for-profit in nature and, in themselves, represent examples of the services segment of the bioscience industries. For-profit, overtly commercial service providers may not see themselves as, or even identify with, biobanks. But conceptually they are biobanks. Users of HBS are often very dependent on commercial biobanks for essential supplies of HBS. The for-profit nature of some suppliers of HBS raises questions about profiteering from trading in parts of the human body and about trafficking in human body parts.[15] These are important considerations for the downstream responsible use of HBS.

---

[13] Peter H. Riegman et al, "Biobanking for better healthcare" (2008) Mol Oncol 2: 213–22.

[14] Robert Hewitt & Peter Watson, "Defining biobank" (2013) Biopreserv Biobank 11: 309–15; Peter H. Watson & Rebecca O. Barnes, "A proposed schema for classifying human research biobanks" (2011) Biopreserv Biobank 9: 327–33.

[15] Lori Andrews & Dorothy Nelkin, *Body Bazaar: The Market for Human Tissue in the Biotechnology Age* (1st edition, Crown 2001); Council of Europe, Convention for the Protection of Human Rights and Dignity of the Human Being with regard to the Application of Biology and Medicine: Convention on Human Rights and Biomedicine [1997] ETS 164, article 21; Council Recommendation CM/Rec(2016)6 of the Committee of Ministers to member States on research on biological materials of human origin [2016] article 6.

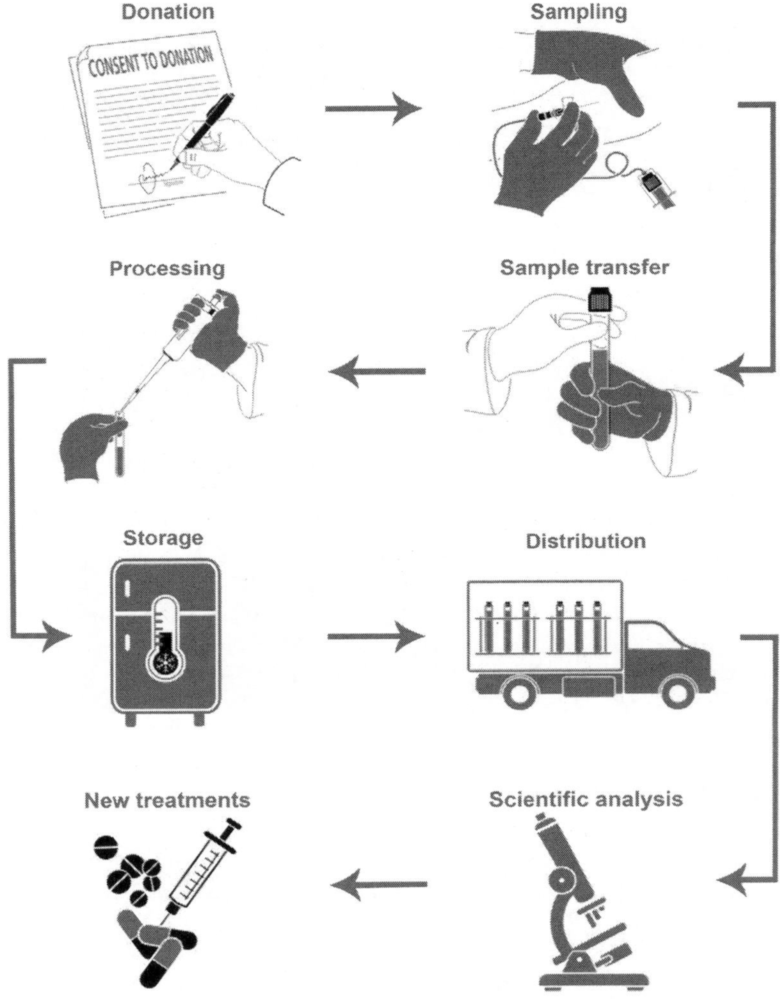

Figure 4.1 *The chain of biobanking processes from donation to end use in research and the development of beneficial products. Responsible governance requires oversight along the entire chain*

Depending on the nature of the HBS, their rarity, the specific characteristics of the donors, and the availability of the HBS, end-users in the bioscience industries may have to seek sources from a variety of organizations, often in several other countries. This adds complexity to the governance requirements as samples originate from, pass through, and may be used in several countries involving several different legal jurisdictions and societal norms. In addition, the supply chains are often multistep with several intermediaries involved, not all of whom are visible to or disclosed to the bioscience company using the HBS (Figure 4.2). Multistep and multinational supply chains multiply the degree of difficulty for a bioscience company to ensure governance processes that effectively secure that their uses of HBS are demonstrably socially responsible, ethical, compliant with applicable laws and regulations, and safe.

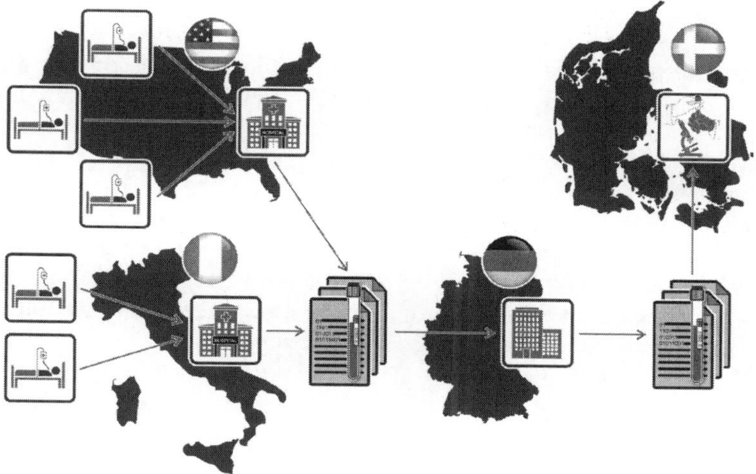

Notes: HBS and associated personal data are collected from donors, at multiple institutions in multiple countries, and routed via a "supplier" in another country, before being supplied to an end user in a further country. The full extent of the supply chain may be invisible to the end-user research organisation, to which it may appear that the HBS originated from the country of the supplier.

Figure 4.2 An example of the complexity of HBS supply chains

4 WHY IS RESPONSIBLE USE OF HUMAN BIOSAMPLES IMPORTANT?

When a part of a human is dissociated from the person from whom it is derived, both physically and temporally, and packaged in a way that closely resembles other laboratory, industrial, or scientific "consumables," it is very easy for users to cease to see the donor behind the HBS—a person who has fundamental rights, hopes, fears, intentions, who is protected by law, and who, at very least, deserves respect.[16] The HBS can come to be seen as a commodity, especially when obtained by a transaction using the language of trading—buying and selling something that can be used for whatever purpose without further thought or restriction.

Without conscious awareness of the special considerations that apply and the necessity to secure that all applicable requirements are met, it is surprisingly easy for users working with HBS to fail to recognize that HBS, or the use of HBS, has special status in ethical, societal, and legal terms in many jurisdictions.[17] The international nature of HBS supply chains also introduces complexity and inconsistency where, for example, HBS are associated with specific legal status and requirements in one country, but not with equivalent legal standards in another. Users also need awareness that international declarations, conventions, consensus statements, guidelines, and standards exist that often extend requirements above the legal minimums, and that these must at least be considered, even if not observed. Table 4.2 lists some of the "hard and soft" requirements documents that apply to the use of HBS and which must be considered when attempting to use HBS responsibly.

These factors taken together generate a perfect storm whereby users may fail to recognize HBS as materials with special status and requirements, may lack awareness of the special requirements which apply, and may not realize the complexity (and inherent risks) in the multistep, multiparty, and multinational supply chains involved. They may show a tendency toward a local view of

---

[16] World Medical Association, "WMA Declaration of Helsinki—ethical principles for medical research involving human subjects" www.wma.net/policies-post/wma-declaration-of-helsinki-ethical-principles-for-medical-research-involving-human-subjects accessed November 20, 2017; Council for International Organizations of Medical Sciences (CIOMS) "International Ethical Guidelines for Health-related Research Involving Humans" https://cioms.ch/wp-content/uploads/2017/01/WEB-CIOMS-EthicalGuidelines.pdf accessed April 23, 2018.

[17] U.S. Department of Health & Human Services, Office for Human Research Protections, "International compilation of human research standards" (2017) www.hhs.gov/ohrp/sites/default/files/international-compilation-of-human-research-standards-2017.pdf accessed October 12, 2017.

Table 4.2 Examples of hard and soft requirements for the use of HBS

Originator	Names or titles of documents
National or international laws, regulations or statutory devices	Various
Council of Europe	Convention on Human Rights and Biomedicine of the Council of Europe (1997)
Council of Europe	Recommendation CM/Rec(2016)6 of the Committee of Ministers to member States on research on biological materials of human origin (2016)
Council for International Organizations of Medical Sciences (CIOMS)	CIOMS Guidelines (2016)
International Conference on Harmonization of Technical Requirements for Registration of Pharmaceuticals for Human Use (ICH)	Good Clinical Practice E6 (R2) (2016)
International Society for Biological and Environmental Repositories (ISBER)	Best Practices for Repositories: Collection, Storage, Retrieval and Distribution of Biological Materials for Research (3rd edition 2012)
National Cancer Institute, National Institutes of Health, U.S. Department of Health and Human Services	NCI Best Practices for Biospecimen Resources (2016)
National Cancer Research Institute, Confederation of Cancer Biobanks (UK)	Biobank Quality Standard: Collecting, Storing and Providing Biological Material and Data for Research (2014)
Organisation for Economic Cooperation and Development (OECD)	OECD Guidelines on Human Biobanks and Genetic Research Databases (2009)
United Nations	Guiding Principles on Business and Human Rights (2011)
World Medical Association	WMA Declaration of Helsinki—Ethical Principles for Medical Research Involving Human Subjects (2013)
World Medical Association	WMA Declaration of Taipei on Ethical Considerations regarding Health Databases and Biobanks (2016)

what is acceptable and allowable that does not factor in the global perspective that is necessary when HBS are obtained from and used in many different countries.

For any user organization, whether a bioscience company, a university, or any other user entity, this is a challenge. Small organizations, such as small biotechnology companies, may not have the necessary resources or financial sustainability for this. Large organizations such as global pharmaceutical companies may have such diverse, dispersed, and complex activities that it is difficult to see the whole picture and effectively manage the situation. However, for organizations of any size, failure to address the issue of responsible use of HBS may adversely affect their legitimacy as responsible users of HBS. Organizational reputation may be impacted, legal consequences may arise, and the social sustainability of the organization may override the financial and resource considerations. Can any bioscience company using HBS afford not to tackle this issue?

5 WHAT ARE THE SOLUTIONS?

There is no single or simple solution. However, there are a few relatively simple measures that bioscience companies and others using HBS can consider to strengthen their approach to the responsible use of HBS. Many organizations in the bioscience industries have adopted their own versions of these suggestions and it is important that any solution is appropriate to the operating context of the individual company. Some companies publish information in the public domain about their approaches to using HBS.[18]

5.1 Embed Responsibility and Commitment at the Level of Top Management

The higher up in the organization the responsibility is embedded, the more likely it is that it will be taken seriously, that the impacts will be properly assessed, and that the necessary resources will be made available to address any issues. This approach is, for good reason, a key requirement in international quality standards.[19]

---

[18] Novo Nordisk A/S, "Human biosamples" www.novonordisk.com/rnd/inside-r-d/bioethics/human-biosamples.html accessed October 13, 2017. Amgen Inc., "Biobanking of human samples" www.amgen.com/about/how-we-operate/policies-practices-and-disclosures/ethical-research/biobanking-of-human-samples/ accessed 13 October 2017. Roche, "Roche position on human specimen resources (biobanks)" www.roche.com/dam/jcr:5aa0f3e7-fa20-4153-a8b8-38df4477712f/10_Position%20on%20Human%20Specimen%20Resources%20(Biobanks)_reviewed_4_2017.pdf accessed October 13, 2017. Pfizer Inc., "Use of human tissue. Use of human biological specimens" www.pfizer.com/research/research_clinical_trials/policy_use_human_tissue accessed October 13, 2017.

[19] International Organization for Standardization, "ISO 9001:2015 Quality management systems—requirements" www.iso.org/standard/62085.html accessed April 23, 2018.

5.2 Foster a Culture where Respect for the Donors of HBS is Paramount

Users of HBS in any organization can be focused on their specific uses of HBS and what it means for their research, projects, products, or sales. They can easily overlook where the HBS has come from and that it is a privilege to use such material—a privilege that comes with duties and responsibilities. If the company is able to foster a culture where staff know that their duties and responsibilities toward the donors are binding and non-negotiable and must never be compromised, this will go a long way to securing responsible use of HBS. This is linked to providing staff and business partners with education about the governance systems and requirements (discussed presently).

5.3 Understand the Laws in Relevant Jurisdictions

It is necessary to develop a means by which the relevant laws and regulations of the jurisdictions where HBS are donated, obtained, and used are well known and by which compliance with them can be assured. The amount of work required to meet this suggestion is dependent on the context for an individual company. It can range from a relatively simple task to one of some complexity. However, no company can afford to operate without cognisance of the laws that apply to its business. It is possible to obtain published overviews[20] and advice from national and international sector specific bodies,[21] in addition to formal legal advice from legal subject matter experts.

5.4 Monitor Developments in International Declarations, Conventions, Consensus Statements, Guidelines, and Standards

It is important, when working across borders, to ensure that the existence and impact of any relevant international declarations, conventions, consensus statements, guidelines, and standards is known. Although many of these documents have no force in law, they may set higher requirements than local laws and represent a harmonized approach that facilitates both compliant interna-

---

[20] U.S. Department of Health & Human Services, *supra* note 16; *see* contents of "Special issue: SYMPOSIUM: Harmonizing privacy laws to enable international biobank research: part I' (2015) J Law Med Ethics 43: 673–826.

[21] Public Population Project in Genomics and Society (P3G) "Data/sample collection—ELSI interoperability" www.p3g.org/datasample-collection-elsi-interoperability_2017 accessed October 3, 2017; BBMRI-ERIC, "Common service ELSI" www.bbmri-eric.eu/BBMRI-ERIC/common-service-elsi/ accessed October 3, 2017.

tional working and best practices. Companies that aim their own approach at or close to the requirements of these international documents are more easily able to work with international partners, suppliers, and customers, as these requirements typically iron out the inconsistencies and incompatibilities introduced by compliance with only local laws. Therefore, a responsible approach is to operate to a "higher common denominator" set of requirements. Keeping up to date with such developments is a relatively easy task as they are highlighted and discussed in many readily available international fora,[22] and updates or changes usually occur after long consultation periods and implementation timelines, allowing those affected to plan for any impacts.

5.5 Governance

An internal governance framework must be in operation that assures the requirement to responsibly use HBS. An effective governance framework may include the following:

- Anchorage of the governance in key management committee(s), such as a company bioethics committee, corporate social responsibility committee or similar. This maintains visibility of the issue and ensures periodic internal review, reporting and linkage to top management.
- Policy commitments, published internally and externally, with alignment to ensure that a holistic approach to policy commitment is taken—human rights, ethics, legal, quality, safety. Depending on the size of the organization and the structure of its governance framework, these aspects could be embedded across a number of company policies or all collected together into a specific policy on the use of HBS.
- Codes of conduct that clearly outline in summary terms what behaviors are expected from employees and business partners, including suppliers.
- Adopting relevant external requirements into internal control processes and procedures. These should detail how employees and business partners must act to comply with the governance requirements.
- Educating employees and business partners about why this is important, as well as on what is expected of them and how they should act to meet the expectations. In many regards this is the most important step as, without an understanding of why and how to comply, governance is difficult to achieve by processes of control alone.

---

22 Public Population Project in Genomics and Society (P3G), *supra* note 31; BBMRI-ERIC, *supra* note 20; International Society for Biological and Environmental Repositories (ISBER), "Science policy resources and tools" www.isber.org/page/SPResources accessed October 13, 2017.

5.6 Monitor: Track, Measure, and Report Internally

A responsible approach also implies a need for accountability. To achieve this, a degree of monitoring is required. An organization using HBS responsibly should consider being able to account for, for example, where the HBS comes from; which types of HBS are used and in what quantities; what the HBS is used for; and what the fate of the HBS will be—are they used to destruction, are they being stored, have they been discarded, have they been transferred to another organization? Reporting such information internally to management committees and top management allows for conscious responsibility and visible commitment.

6 KEY PROCESSES FOR SPECIAL ATTENTION

The solutions just suggested may seem onerous or even difficult to achieve in some cases. However, if a few key processes are adequately controlled, most of what is required to fulfil a responsible approach to the use of HBS can be met. Organizations controlling where they obtain HBS from, how they store HBS and associated data, where they transfer HBS to, how they approve and monitor uses of HBS, and how all of this is recorded can demonstrate a responsible approach.

6.1 Control Sourcing of HBS

The legitimacy of any use of HBS starts with the originator—the donor. That legitimacy must be preserved throughout the supply chain to be available to the end user.

Consent is the means by which donors express, or agree to, the terms under which their HBS can be taken, processed, stored, transferred, and used. An in depth discussion about the legal and ethical technicalities of valid and informed consent is beyond the scope of this chapter.[23] However, responsible bioscience companies, or any user of HBS, should consider, develop, and implement minimum standards for the content of consent documents and the means by which consent is obtained, as part of their governance standards.

---

[23] Timothy Caulfield & Blake Murdoch, "Genes, cells, and biobanks: yes, there's still a consent problem" (2017) PLoS Biol 15:e2002654; Palmira Granados Moreno & Yann Joly, "Informed consent in international normative texts and biobanking policies: seeking the boundaries of broad consent" (2016) Medical Law International 1–30; Kristin Solum Steinsbekk, Bjørn Kåre Myskja, & Berge Solberg, "Broad consent versus dynamic consent in biobank research: is passive participation an ethical problem?" (2013) European Journal of Human Genetics 21: 897–902.

These standards should at least meet the legal requirements of all relevant jurisdictions in the countries where HBS are donated, stored, and used, and wherever possible meet international consensus recommendations in order to secure the ability to operate internationally.[24] It is very important that the scope of the consent includes the intended use. It is desirable that consent documents are transparent in informing their donors that commercial companies may be users,[25] and that their HBS may be moved internationally, if relevant. Those who supply HBS to bioscience companies should be asked to demonstrate that legally valid consent from donors exists, in a form that meets the companies' standards.

Human biosamples are rarely donated, supplied, or used without some accompanying annotating information pertaining to donors. Debates exist about the myth or reality of true anonymity,[26] so some presumption of the theoretical possibility that accompanying personal data may become identifiable is recommended. Therefore, donor consent should also cover the acquisition, processing, storage, transfer, and usage of sensitive personal data, even if considered to be irreversibly anonymous.

Legitimacy for use of HBS also comes from any approvals, authorizations, restrictions, or similar that official or competent government bodies may give to the sources of HBS. Bioscience companies need to check that these approvals allow for the transfer of the HBS to them and for their intended uses. Contractual obligations, freedoms, and restrictions that existed along the supply chain prior to receipt by the biosample company also need to be checked, considered, and agreed.

Other aspects to confirm when sourcing HBS are as follows:

- There must be a clear, documented, and traceable chain of custody back along the supply chain to the point of donation. This is the only way for a user to confidently confirm the origin of HBS, the consent and approvals that apply, and the overall *bona fides* of the HBS. On occasions, parties in the supply chain aim to conceal this information for various reasons.

---

24 World Medical Association, "WMA Declaration of Taipei on ethical considerations regarding health databases and biobanks" (World Medical Association, 2016) www.wma.net/policies-post/wma-declaration-of-taipei-on-ethical-considerations-regarding-health-databases-and-biobanks/ accessed April 23, 2018.

25 Timothy Caulfield, Pascal Borry, & Herbert Gottweis, "Industry involvement in publicly funded biobanks" (2014) Nature Review Genetics 15: 220.

26 Jane Kaye, "The tension between data sharing and the protection of privacy in genomics research" (2012) Annu Rev Genomics Hum Genet 13: 415–31; Adrian Thorogood & Ma'n H Zawati, "International guidelines for privacy in genomic biobanking (or the unexpected virtue of pluralism)" (2015) J Law Med Ethics 43: 690–702.

However, failure to disclose this information can impair trust between organizations and mask real issues that have occurred intentionally, maliciously, by accident, or by omission. A responsible user should seek to ensure that there are no hidden risks or secrets relating to the origins and supply chain that could impact the legitimate and responsible use of HBS.

- The staff of the supplier must work in a safe and healthy environment. This latter consideration is a key requirement to ensure the respect for human rights that a responsible bioscience company should maintain. Unsafe, exploitative, or negligent working conditions for the staff of a supplier are issues for companies receiving goods or services from the supplier, not just for the supplier itself.[27]
- The organization supplying HBS must be sufficiently skilled, expert where necessary, and equipped for the task.
- The quality of the HBS must match expectations, the supplier must have implemented all necessary processes and procedures to assure quality, and the HBS supplied must be fit for purpose. Quality of HBS can be adversely impacted during collection, processing, storage, and shipping before use. Poor quality HBS results in poor quality or misleading or dangerous uses—each of which can become an ethical, legal, or safety issue.

6.2 Storage of HBS Must be Controlled, Including Storage of Associated Data

Human biosamples are precious and often rare or difficult to obtain. They also contain sensitive personal information about the donors, such as genetic information or information about health or disease status. As such, HBS need to be stored in ways that preserve their integrity and utility and access must be controlled to protect against unwarranted use, damage, loss, theft, or malicious tampering. All of these factors apply as much to data (paper, electronic, or otherwise) that accompany HBS, are derived from HBS, or are otherwise associated with HBS, especially if the HBS and data have not been rendered irreversibly anonymous.

6.3 Movement Must be Controlled

Human biosamples are generally regarded as having the potential to be biohazardous. They can be known to be infectious, be suspected to be infectious,

---

[27] United Nations, "Guiding principles on business and human rights" (United Nations, 2011) www.ohchr.org/Documents/Publications/GuidingPrinciplesBusinessHR_EN.pdf accessed April 23, 2018.

or have an infectious potential regarded as low but essentially unknown, as it is virtually impossible to exclude the presence of all potential pathogens by testing and clinical likelihood alone.

Except when the HBS is a therapeutic product, for which special requirements apply, it is necessary to package and transport HBS for other uses according to national and international dangerous goods regulations.[28] This generally means classifying the HBS as infectious or potentially infectious and using packaging, labelling, and shipping arrangements that are compliant with the regulations.

6.3 Uses of HBS Must be Controlled

Possession of HBS does not provide *carte blanche* to use them in any way desired. Legitimacy of use requires that the use meets all necessary legal and regulatory requirements, that it is safe, it is in accordance with the freedoms and restrictions imposed by donor consent, approvals by competent bodies, and contracts with upstream suppliers. Uses must also be safe and of sufficient quality to be fit for purpose. A responsible approach requires that organizations have internal processes for reviewing and approving activities using HBS.

The degree of difficulty involved in implementing and operating these key processes will vary from organization to organization, but control of complex sourcing arrangements is probably the most onerous issue for most bioscience companies.

7 FINAL THOUGHTS

Members of the bioscience industries are legitimate users of HBS in their pursuit of the discovery, development, manufacturing, and provision of safe and effective products and services that deliver benefits for patients and society. Legitimacy is not enough, however, and responsible bioscience industries must care about the governance of using human biosamples. Responsible use of human biosamples requires compliance with bioethics norms, laws and

---

[28] Economic Commission for Europe, Committee on Inland Transport, "ADR 2017: European agreement concerning the international carriage of dangerous goods by road" (UNECE, 2017) www.unece.org/trans/danger/publi/adr/adr2017/17contentse0.html accessed April 23, 2018; International Air Transport Association (IATA), "2018 Dangerous Goods Regulations (DGR)" (IATA, 2017) www.iata.org/publications/dgr/Pages/index.aspx accessed 23 April 2018; International Civil Aviation Organization (ICAO), "Technical instructions for the safe transport of dangerous goods by air" (ICAO, 2017) www.icao.int/safety/DangerousGoods/Pages/technical-instructions.aspx accessed April 23, 2018.

regulations, societal expectations, and applicable international declarations, conventions, standards, and best practices. Striving for a highest common denominator standard more readily facilitates international sourcing and use of HBS.

Good governance "starts at home"—internal to the company. But this is not enough. It must also extend to suppliers, business partners, and collaborators, as each have the potential by accident, intent, malice, negligence, or omission to fail to reach the standards required to assure a responsible approach to using HBS. This is challenging for any organization, but it is achievable even in complex global companies.

Public trust and confidence in industrial and commercial users of HBS, as legitimate and responsible users, needs to be won and maintained. Ultimately, this requires transparency, communication, and engagement to demonstrate that individual companies and the industries as a whole are taking a responsible approach.[29]

Of course, the considerations outlined in this chapter are not unique to bioscience industries. The same or similar issues can equally apply to any users of HBS in public or nonprofit institutions, such as universities. For these other organizations, similar solutions may be considered appropriate to demonstrate that they are legitimate and responsible users of HBS.

---

[29] Novo Nordisk A/S, "Human biosamples in pharmaceutical research: a responsible approach" www.novonordisk.com/content/dam/Denmark/HQ/RND/Documents/Novo%20Nordisk_Human%20Biosamples%20Brochure_Aug%202017_web.pdf.

PART III

The interface of biobanks, human rights, and patient involvement

5. Biobanking, scientific productions and human rights

Peter K. Yu[1]

1 INTRODUCTION

Biobanks exist in many forms, sizes, designs, and structures.[2] They provide important benefits to humanity.[3] For instance, biobanks can help to predict genetically inherited diseases and develop diagnostics or therapeutics to address them. They can also help to reduce treatment costs, increase healthcare options, issue lifestyle advice, and facilitate the development of preventive measures. They can even advance the field of biomedical and genomic research while sparking breakthroughs in the area of personalized medicines.

Notwithstanding these important benefits, the issue of biobanks has raised serious concerns and complications, especially in the areas of privacy, autonomy, and personal data protection. Because these concerns often implicate the rights of individual donors, biobank users, relevant family members, and other

---

[1] This chapter draws on research from the author's earlier article in the *SMU Law Review* and a book chapter published by Cambridge University Press.

[2] Herbert Gottweis and Alan Petersen, "Biobanks and Governance: An Introduction" in Herbert Gottweis and Alan Petersen (eds), *Biobanks: Governance in Comparative Perspective* (Milton Park: Routledge 2008) 5; Michaela Mayrhofer, "Patient Organizations as the (Un)usual Suspects: The Biobanking Activities of the Association Française Contre Les Myopathies and Its Généthon DNA and Cell Bank" in Gottweis and Petersen, *Biobanks* 71–2; Darren Shickle and Marcus Griffin, "Biobanks, Networks and Networks of Networks" in Kris Dierickx and Pascal Borry (eds), *New Challenges for Biobanks: Ethics, Law and Governance* (Antwerp: Intersentia 2009) 2; Mark Stranger and Jane Kaye, "Governing Biobanks: An Introduction" in Jane Kaye and Mark Stranger (eds), *Principles and Practice in Biobank Governance* (Farnham: Ashgate Publishing 2009) 2.

[3] Gottweis and Petersen, *supra* note 2, at 3, 27–8; Atieh Zarabzadeh et al, "Ensuring Participant Privacy in Networked Biobanks" in Kaye and Stranger, *supra* note 2, at 178.

individual third parties, a logical topic to explore is whether a violation of the rights of these individuals could rise to the level of a human right violation.

Although the scope and length of this chapter do not allow for a full exploration of all the different human rights issues involved in the area of biobanking, it aims to provide a brief survey on three distinct sets of issues, all related to biobanks. The first set concerns the human rights involved in the collection, processing, use, or storage of the biological materials collected by biobanks. The second set pertains to the human rights issues implicated by the development of scientific productions utilizing the collected materials. The third set relates to the human rights obligations of three types of biobanks: public biobanks, private biobanks, and biobanks formed out of public–private partnerships (PPPs).

The goal of this chapter is not to provide detailed analyses of these three sets of human rights issues. Instead, it aims to offer preliminary sketches of the various human rights issues that can be implicated by biobanks. It is my hope that this chapter will highlight the complexities concerning human rights issues in the area of biobanking and thereby generate greater interest and attention from the medical community, biobank operators and funders, policymakers, regulators, commentators, and the mass media.

2 HUMAN RIGHTS FRAMEWORK

The first set of human rights issues concerns the collection, processing, use, or storage of the biological materials collected by biobanks. These issues are the easiest to explore, thanks to the large number of international and regional human rights documents that have already been adopted in the area of biobanking.

As far as human rights are concerned, the oft-used starting points are the Universal Declaration of Human Rights (UDHR) and the International Covenant on Civil and Political Rights (ICCPR). Article 12 of the UDHR states: "No one shall be subjected to arbitrary interference with his privacy, family, home or correspondence, nor to attacks upon his honour and reputation. Everyone has the right to the protection of the law against such interference or attacks." Although this provision covers a wide array of privacy issues—both related and unrelated to biobanking—it is worth recalling that the drafters of the UDHR were deeply disturbed by the abuse of science and technology during World War II.[4]

---

4 Audrey Chapman, "A Human Rights Perspective on Intellectual Property, Scientific Progress, and Access to the Benefits of Science" in *Intellectual Property and Human Rights* (Geneva: World Intellectual Property Organization 1998) 131; Richard

When the UDHR language was subsequently transposed onto the ICCPR to create a legally binding covenant, article 17 of that covenant incorporated language that was virtually identical to article 12 of the UDHR. Article 1 of the ICCPR also makes clear that "[a]ll peoples have the right of self-determination." Such a right is particularly important in the context of biobanking, whether one focuses on individual privacy, personal autonomy, informed consent, or data misuse.

In the biomedical field, the leading international human rights instrument is the Convention for the Protection of Human Rights and Dignity of the Human Being with Regard to the Application of Biology and Medicine.[5] Known widely as the Oviedo Convention or the Convention on Human Rights and Biomedicine, this instrument was adopted by the Council of Europe in Oviedo, Spain in April 1997. Because this document remains the only major international human rights instrument in this area, this section discusses its provisions in greater length.

Article 1 of the Oviedo Convention requires each contracting party to "protect the dignity and identity of all human beings and [to] guarantee everyone, without discrimination, respect for their integrity and other rights and fundamental freedoms with regard to the application of biology and medicine." Article 2, which carries a heading of "primacy of the human being," states that "[t]he interests and welfare of the human being shall prevail over the sole interest of society or science."

Chapter II focuses on the donors' informed consent, an issue that is of vital importance in the governance of biobanks. Article 5 stipulates:

> An intervention in the health field may only be carried out after the person concerned has given free and informed consent to it.
>
> This person shall beforehand be given appropriate information as to the purpose and nature of the intervention as well as on its consequences and risks.
>
> The person concerned may freely withdraw consent at any time.

While it is easy to understand the need for informed consent and the withdrawal of consent after it has been granted, giving individuals the ability to withdraw consent *at any time* can present major challenges to biobanks (as well as to

Pierre Claude, "Scientists' Rights and the Human Right to the Benefits of Science" in Audrey Chapman and Sage Russell (eds), *Core Obligations: Building a Framework for Economic, Social and Cultural Rights* (Antwerp: Intersentia 2002) 249–50.

5 Convention for the Protection of Human Rights and Dignity of the Human Being with Regard to the Application of Biology and Medicine, 36 ILM 817 (adopted April 4, 1997).

any downstream developer of scientific productions utilizing the biological materials these banks collect). To begin with, the success of biobanks and their collaborators often depends on scale and the banks' ability to aggregate the collected data and integrate them with those found in other biobanks.[6] Such aggregation and integration are particularly needed when big data analyses are deployed or when the banks are part of large, often international networks.[7]

Moreover, complications concerning the information needed to obtain informed consent will inevitably arise.[8] As Geraldine Fobelets and Herman Nys have observed:

> [W]hen participants consent, specific research purposes are often not yet known at the moment of collection and as a result, unexpected findings may be made … [E]ven when research purposes are known, unexpected findings can still be made, since "DNA contains an individual's probabilistic future diary, written in a code that has only partially been broken". Every further deciphering of the code may therefore lead to new and unexpected information concerning the participants.[9]

If the participants' withdrawal requires the destruction of key data or tissue samples that are critical to the development of diagnostics or therapeutics, as opposed to merely rendering these data or samples nonidentifiable,[10] such destruction could raise competing human rights claims based on the right to

---

6 Council of Europe, "Recommendation Rec(2006)4 of the Committee of Ministers to Member States on Research on Biological Materials of Human Origin," explanatory memorandum, para 52; Shickle and Griffin, *supra* note 2, at 1; Stranger and Kaye, *supra* note 2, at 2.

7 Gottweis and Petersen, *supra* note 2, at 6.

8 As the Explanatory Report to the Oviedo Convention states, "In order for their consent to be valid the persons in question must have been informed about the relevant facts regarding the intervention being contemplated. This information must include the purpose, nature and consequences of the intervention and the risks involved. Information on the risks involved in the intervention or in alternative courses of action must cover not only the risks inherent in the type of intervention contemplated, but also any risks related to the individual characteristics of each patient, such as age or the existence of other pathologies. Requests for additional information made by patients must be adequately answered." Council of Europe, "Explanatory Report to the Convention for the Protection of Human Rights and Dignity of the Human Being with Regard to the Application of Biology and Medicine: Convention on Human Rights and Biomedicine" (1997) para 35 ("Explanatory Report").

9 Geraldine Fobelets and Herman Nys, "Evolution in Research Biobanks and Its Legal Consequences" in Dierickx and Borry, *supra* note 2, at 24.

10 Article 13(1) of the Council of Europe's 2016 Recommendation on Research on Biological Materials of Human Origin stipulates, "When identifiable biological materials are stored for research purposes only, the person who has withdrawn consent should have the right to have, in the manner foreseen by law, the materials and associated data either destroyed or rendered non-identifiable."

life, the right to health, or even the right to "enjoy the benefits of scientific progress and its applications," which has remained obscure until recently.[11]

Chapter III of the Oviedo Convention turns to the right to information, another important issue in the area of biobanking. Article 10(1) states that "[e]veryone has the right to respect for private life in relation to information about his or her health." Article 10(2) further provides: "Everyone is entitled to know any information collected about his or her health. However, the wishes of individuals not to be so informed shall be observed."

Like the ability to withdraw consent at any time, the issue of an individual's wish not to be informed has presented major challenges to biobanks. While these banks could make a conscious choice to withhold information from those who have exercised their right not to know, indirect communication is often difficult to avoid. Indeed, there may be difficult situations in which biobanks have to juggle a donor's right not to be informed with the right to information of his or her family members or other relevant third parties. As the explanatory report to the Oviedo Convention states:

> [C]ertain facts concerning the health of a person who has expressed a wish not to be told about them may be of special interest to a third party, as in the case of a disease or a particular condition transmissible to others, for example. In such a case, the possibility for prevention of the risk to the third party might ... warrant his or her right taking precedence over the patient's right to privacy ... and as a result the right not to know ... In any case, the right not to know of the person concerned may be opposed to the interest to be informed of another person and the interests of these two persons should be balanced by internal law.[12]

Similarly, Lynn Dressler has reminded us: "In the legal setting duty to warn cases have involved the determination of whether or not the physician was

---

[11] On this right, see Aurora Plomer, *Patents, Human Rights and Access to Science* (Cheltenham: Edward Elgar Publishing 2015); UNESCO, *The Right to Enjoy the Benefits of Scientific Progress and Its Applications* (Geneva 2009); Margaret Weigers Vitullo and Jessica Wyndham (eds), *Defining the Right to Enjoy the Benefits of Scientific Progress and Its Applications: American Scientists' Perspectives* (Washington: American Association for the Advancement of Science 2013); Lea Shaver, "The Right to Science and Culture" (2010) Wisconsin L Rev 121; William A. Schabas, "Study of the Right to Enjoy the Benefits of Scientific and Technological Progress and Its Application" in Yvonne Donders and Vladimir Volodin (eds), *Human Rights in Education, Science and Culture: Legal Developments and Challenges* (Aldershot: Ashgate Publishing 2007).

[12] See "Explanatory Report," *supra* note 8, para 70.

responsible to warn a family member of their risk of harm, based on knowledge [of a mutation, for example] of the individual the physician is treating."[13]

Chapter IV covers nondiscrimination and nonstigmatization, which are important considering that sensitive genetically related information can fall into the hands of employers and insurers.[14] Article 11 prohibits "[a]ny form of discrimination against a person on grounds of his or her genetic heritage." Articles 12 and 14 lay down strict restrictions concerning different activities, including the tests that can be used to predict genetic diseases or detect predispositions or susceptibilities to those diseases, interventions on the human genome, and the use of medically assisted procreation techniques to choose the sex of future children.

Apart from the Oviedo Convention, several other international and regional documents have provided useful information and normative language concerning ways to strengthen human rights protection in the area of biobanking. The first document is the Universal Declaration on the Human Genome and Human Rights, which the United Nations Educational, Scientific and Cultural Organization (UNESCO) adopted in November 1997.[15] Article 2(a) states explicitly that "[e]veryone has a right to respect for their dignity and for their rights regardless of their genetic characteristics." Article 5(c) further calls for respect for "the right of each individual to decide whether or not to be informed of the results of genetic examination and the resulting consequences should be respected." In addition, article 6 states that "[n]o one shall be subjected to discrimination based on genetic characteristics that is intended to infringe or has the effect of infringing human rights, fundamental freedoms and human dignity."

The second document is the *OECD Guidelines on Human Biobanks and Genetic Research Databases*, which the Organisation for Economic Co-operation and Development (OECD) released in October 2009.[16] Principle 1.D states that "the operators and users of the HBGRD [human biobank and genetic research database] should respect human rights and freedoms and secure the protection of participants' privacy and the confidentiality of data

---

13 Lynn G. Dressler, "Biobanking and Disclosure of Research Results: Addressing the Tension Between Professional Boundaries and Moral Intuition" in Jan Helge Solbakk, Søren Holm and Bjørn Hofmann (eds), *The Ethics of Research Biobanking* (Dordrecht: Springer 2009) 92.

14 Fobelets and Nys, *supra* note 9, at 25.

15 In addition to this document, UNESCO adopted the International Declaration on Human Genetic Data (adopted October 16, 2003) and the Universal Declaration on Bioethics and Human Rights (adopted October 19, 2005).

16 Organisation for Economic Co-operation and Development, *OECD Guidelines on Human Biobanks and Genetic Research Databases* (Paris 2009).

and information." Principle 1.F states further that these operators "should develop and maintain clearly documented operating procedures and policies for the procurement, collection, labelling, registration, processing, storage, tracking, retrieval, transfer, use and destruction of human biological materials, data and/or information."

Also included in the OECD Guidelines are provisions concerning the commercial use of biological materials collected through HBGRDs, an issue that has important implications on the protection of intellectual property rights. Principle 9.B states that "[b]enefits arising from research using the HBGRD's resources should be shared as broadly as possible, including by the sharing of information, licensing, or transferring of technology or materials." Principle 9.C further states that "[t]he operators of the HBGRD should have a clearly articulated policy and explicitly indicate to participants whether they and/or the HBGRD retain any rights over the human biological materials and/or data and the nature of such rights." Principle 9.D calls on these operators to "have a clearly articulated policy that is communicated to participants relating to the commercialization of its own resources, research results derived from those resources, and/or commercial products, if any, that may arise from research using its resources."

The final document is the Council of Europe's latest recommendation on research on biological materials of human origin.[17] Adopted in May 2016 to update the recommendation released a decade ago, this new recommendation built on the Oviedo Convention, also adopted by the Council of Europe. Specifically, the recommendation contains an extensive section covering the governance of collections of biological materials, including biobanks. Articles 16 to 20 include provisions on governance principles, feedback of health-related information, researchers' access to collected biological materials, the transborder flows of these materials, and oversight provided to the covered collections. The recommendation also includes a section on the use of the collected biological materials in research projects, addressing issues relating to commercialization, ownership, benefit sharing, and intellectual property rights.

In sum, many international and regional documents exist to strengthen human rights protection in the area of biobanking. Of particular importance is the respect for privacy and autonomy, informed consent and the ability to withdraw such consent, the right to information and not to receive information, and the protection against discrimination and stigmatization based on genet-

---

[17] Recommendation CM/Rec(2016)6 of the Committee of Ministers to Member States on Research on Biological Materials of Human Origin (adopted May 11, 2016).

ically related information derived from the biological materials collected by biobanks.

3 SCIENTIFIC PRODUCTIONS

The second set of human rights issues pertains to the development of scientific productions utilizing the biological materials collected by biobanks. Compared with the human rights issues explored in the previous section, the issues involved here are much more complicated. They involve not only biobanks, their donors and users, and related family members, but also third parties who have made considerable investment in time, effort, and resources to develop scientific productions utilizing the biological materials collected by biobanks.

Such productions usually attract protection through various forms of intellectual property rights, including patents, trade secrets, and the protection of other undisclosed information. While intellectual property rights are not protected at the same level of human rights, some aspects of the former have been recognized by international and regional human rights instruments. For instance, article 27(2) of the UDHR states expressly that "[e]veryone has the right to the protection of the moral and material interests resulting from any scientific, literary or artistic production of which he [or she] is the author." Article 15(1)(c) of the International Covenant on Economic, Social and Cultural Rights (ICESCR) further requires each state party to the Covenant to "recognize the right of everyone … [t]o benefit from the protection of the moral and material interests resulting from any scientific, literary or artistic production of which he [or she] is the author."

To further complicate matters, Protocol No 1 to the European Convention of Human Rights offers human rights-like protection to corporate intellectual property rights holders.[18] Article 1 of the Protocol specifically provides: "Every natural or legal person is entitled to the peaceful enjoyment of his possessions. No one shall be deprived of his possessions except in the public interest and subject to the conditions provided for by law and by the general principles of international law." The "peaceful enjoyment of … possessions" language is later interpreted by the Grand Chamber of the European Court of Human Rights, in *Anheuser-Busch, Inc v Portugal*, to cover both registered trademarks and trademark applications of a multinational corporation.[19] The protection of corporate intellectual property rights holders has been further

[18] Protocol to the Convention for the Protection of Human Rights and Fundamental Freedoms, 213 UNTS 262 (opened for signature March 20, 1952).

[19] 73049/01 *Anheuser-Busch, Inc v Portugal* ECLI:CE:ECHR:2007:0111JUD007304901.

reinforced by the "[i]ntellectual property shall be protected" language in article 17(2) of the Charter of Fundamental Rights of the European Union, which entered into force in December 2009.[20]

In light of the recognition in these international and regional human rights instruments, tension, or even conflict, may arise between the human rights discussed in the previous section and the right to the protection of the moral and material interests resulting from the scientific productions utilizing the biological materials collected by biobanks. Initially, policymakers and commentators engaged in heated debate over whether the conflict approach or the coexistence approach should be used to address issues lying at the intersection of intellectual property and human rights. By now, however, most commentators have abandoned the conflict approach,[21] considering the difficulty in reconciling such an approach with the express recognition of the right to the protection of the interests resulting from intellectual productions in the UDHR, the ICESCR, and other international or regional human rights instruments. That human rights are "universal, indivisible and interdependent and interrelated," as stated in the Vienna Declaration and Programme of Action, has made it even more difficult for one to argue that the human rights covered in the previous section are more fundamental than the right to the protection of the interests resulting from intellectual productions.[22]

In previous works, I have noted the need to separate the conflicts between human rights and intellectual property rights into two sets of conflicts: external and internal.[23] While internal conflicts exist only within the human rights regime, thereby requiring approaches to resolve competing fundamental rights, external conflicts suggest that intellectual property rights have expanded beyond the requirements of international or regional human rights instruments. Thus, in the event of a conflict between intellectual property and human rights, one can apply the principle of human rights primacy. Endorsed by the UN Sub-Committee on the Promotion and Protection of Human Rights

---

[20] Charter of Fundamental Rights of the European Union [2000] OJ C 364/1, art. 17(2).

[21] On the conflict or coexistence approach, see Laurence R. Helfer, "Human Rights and Intellectual Property: Conflict or Coexistence?" (2003) 5 Minnesota Intellectual Property Rev 47, 48–9; Peter K. Yu, "Ten Common Questions about Intellectual Property and Human Rights" (2007) 23 Georgia State U L Rev 709, 709–11.

[22] "Vienna Declaration and Programme of Action," para 5, (1993) A/CONF.157/23.

[23] Peter K. Yu, "Intellectual Property and Human Rights in the Nonmultilateral Era" (2012) 64 Florida L Rev 1045, 1091–6; Peter K. Yu, "Reconceptualizing Intellectual Property Interests in a Human Rights Framework" (2007) 40 UC Davis L Rev 1039, 1075–123.

in Resolution 2000/7, this primacy principle subordinates the non-human rights aspects of intellectual property rights to human rights obligations.[24]

Building on these earlier works, I further explored in a recent article the complex interactions among scientific productions, intellectual property, and human rights. To help clarify the different types of interests and protections within a human rights framework for intellectual property, I organized the framework into four structural layers (see Figure 5.1). Such organization comes in handy when we seek to pinpoint the specific human rights interests involved in the scientific productions utilizing the biological materials collected by biobanks.

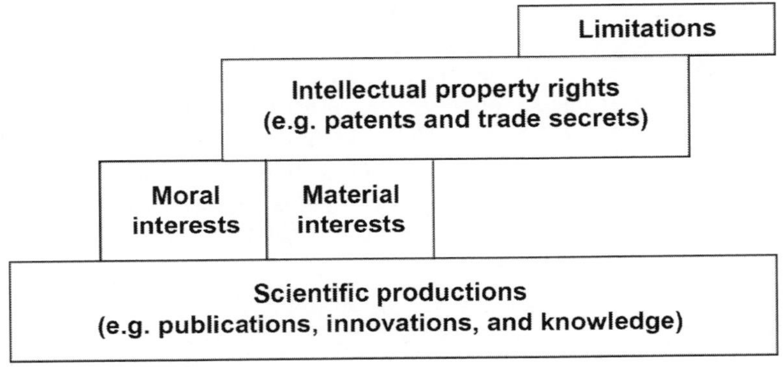

Figure 5.1 *The structure of a human rights framework for intellectual property*

At the bottom of my proposed hierarchical structure is the production layer. As far as scientific productions are concerned, this layer covers all types of scientific productions, including scientific publications, scientific innovations (such as inventions), and scientific knowledge.[25] Many would also include indigenous knowledge, innovations, and practices in this list, although such

24 Sub-Commission on Human Rights, "Intellectual Property Rights and Human Rights" paras 2–3, (2000) E/CN.4/Sub.2/RES/2000/7; Yu, "Reconceptualizing Intellectual Property Interests," *supra* note 23, at 1092–3.

25 Committee on Economic, Social and Cultural Rights, "General Comment No. 17: The Right of Everyone to Benefit from the Protection of the Moral and Material Interests Resulting from Any Scientific, Literary or Artistic Production of Which He or She Is the Author (Article 15, Paragraph 1(c), of the Covenant)" para 9, (2006) E/C.12/GC/17.

inclusion is far from uncontroversial. Undoubtedly, all productions lie in this particular layer. Nevertheless, it will still be important to identify the specific types of productions involved, because each type of production may require different forms of protection. Due to eligibility requirements in intellectual property laws, some of these productions may also need protection outside the existing intellectual property regime.

Above the production layer is the interest layer. This layer covers two different types of human rights interests resulting from intellectual productions: moral and material.[26] While the laws of most countries offer strong protection to material interests—due in large part to the requirements of the Agreement on Trade-Related Aspects of Intellectual Property Rights (TRIPS Agreement) of the World Trade Organization (WTO) and other intellectual property instruments administered by the World Intellectual Property Organization—these laws offer very limited protection, if any, to moral interests. A case in point is the patent system in the United States and many other jurisdictions. This system does not protect the integrity of a patented invention.[27] Nor does it recognize the moral and material interests of subsequent inventors once a patent has issued to the first inventor.[28]

Above the interest layer is the protection layer, which covers the different forms of protection that can be used to address the moral and material interests resulting from intellectual productions. This layer includes not only different intellectual property rights but also protections outside the intellectual property regime, such as those provided through contracts, investment law, food and drug regulations, or alternative funding models.[29]

The top layer is the limitation layer, which covers the different limitations on the rights covered in the protection layer. Just because the intellectual property system has offered strong protection to certain rights does not mean that the human rights framework will protect those rights to the fullest extent. Moreover, one should not forget that intellectual property rights are limited by both endogenous and exogenous limits.[30] Because the interest layer has already internalized the endogenous limits within the covered rights, the limitation layer includes only exogenous limits. Examples of these exogenous limits are

---

[26] Peter K. Yu, "The Anatomy of the Human Rights Framework for Intellectual Property" (2016) 69 SMU L Rev 37, 55–6.

[27] *Ibid* at 50–1; Justin Hughes, "The Philosophy of Intellectual Property" (1988) 77 Georgetown LJ 287, 351.

[28] Robert Nozick, *Anarchy, State, and Utopia* (New York: Basic Books 1974) 182; Yu, *supra* note 26, at 49–50.

[29] Yu, *supra* note 26, at 62–3.

[30] Peter K. Yu, "The Political Economy of Data Protection" (2010) 84 Chicago-Kent L Rev 777, 794–6.

those found in human rights treaties, constitutions, and competition law, or in relation to "morality, public order and the general welfare in a democratic society," as provided in article 29(2) of the UDHR.[31]

This limitation layer can be quite important in the area of biobanking. Because scientific innovations inherently cover scientific knowledge, considerable attention will have to be paid to those limits that aim to prevent the overprotection of intellectual property rights. In the United States, these limits include the idea–expression dichotomy in copyright law, the nonprotection of "laws of nature, natural phenomena, and abstract ideas" in patent law, and the ability to reverse engineer in trade secret law.

Similar exogenous limits can be found at the international level. Although the Paris Convention for the Protection of Industrial Property does not define what constitutes an invention,[32] article 27.2 of the TRIPS Agreement, which incorporated the Convention by reference, allows WTO member states to "exclude from patentability inventions, the prevention … of the commercial exploitation of which is necessary to protect *ordre public* or morality." Article 27.3 further allows these states to exclude from patentability "diagnostic, therapeutic and surgical methods for the treatment of humans or animals" as well as "essentially biological processes for the production of plants or animals other than non-biological and microbiological processes."

To be sure, the hierarchical structure advanced in this section is predicated on a *positivist* reading of existing international human rights instruments, such as the UDHR and the ICESCR. Nevertheless, the structure remains relevant even if we conceptualize the human rights framework for intellectual property differently—such as through first principles, natural law, or moral philosophy. Under such a reconceptualization, the structure and its underlying layers may change. Even if some or all of the previously identified layers remain, they may also interact with each other differently. After all, there is no requirement that the hierarchical structure has four distinct layers—namely, production, interest, protection, and limitation. Nor is there any necessary relationship between each layer.

In sum, regardless of whether one subscribes to a positivist or philosophical conception of the human rights framework for intellectual property, the hierarchical structure advanced in this section can be used to enhance our understanding of the human rights interests involved in the scientific productions utilizing the biological materials collected by biobanks. Indeed,

---

[31] Yu, *supra* note 26, at 65–6.

[32] Daniel Gervais, "Human Rights and the Philosophical Foundations of Intellectual Property" in Christophe Geiger (ed.), *Research Handbook on Human Rights and Intellectual Property* (Cheltenham: Edward Elgar Publishing 2015) 89, 91.

a close scrutiny of the interactions between the different layers can provide important insight into the interplay of intellectual property and human rights. For instance, a top-down focus on the interaction between the protection layer and the lower interest layer shows how the levels of protection in the current intellectual property regime have exceeded the requirements of international and regional human rights treaties. By contrast, a bottom-up focus on the interaction between the interest layer and the upper protection layer helps us locate the appropriate protection for the specific human rights interests identified in the interest layer, or determine whether alternative forms of protection exist to secure those interests.

4 BIOBANKS

The last set of human rights issues relates to the human rights obligations of biobanks. Because biobanks exist in many forms, sizes, designs, and structures, this section separates them into three broad categories: (1) public biobanks; (2) private biobanks; and (3) biobanks formed out of PPPs. The section discusses in turn the human rights obligations of each type of biobank.

4.1 Public Biobanks

With respect to public biobanks, the analysis of their human rights obligations is the most straightforward. Because international human rights instruments tend to be drafted with public actors in mind, these instruments can be easily used to identify and clarify the human rights obligations of public biobanks. As noted in the first section, biobanking will implicate human rights issues concerning the respect for privacy and autonomy, informed consent and the ability to withdraw such consent, the right to information and not to receive information, and protection against discrimination and stigmatization based on genetically related information derived from the biological materials collected by biobanks. Public biobanks will therefore take on obligations relating to these issues as specified in international and regional human rights instruments.

4.2 Private Biobanks

With respect to private biobanks, the analysis of their human rights obligations is much more complicated. Because private actors are generally not parties to international human rights instruments, they do not have the same obligations as public biobanks.

The most authoritative document outlining the human rights responsibilities of private actors is the Guiding Principles on Business and Human Rights

(Guiding Principles),[33] developed by John Ruggie in his capacity as the UN Secretary-General's Special Representative on the Issue of Human Rights and Transnational Corporations and Other Business Enterprises. This document applies to not only States but also transnational corporations, not-for-profit organizations and other private actors.

The origin of the Guiding Principles can be traced back to the "protect, respect and remedy" framework Ruggie delivered to the Human Rights Council in April 2008. This framework rests on three distinct pillars:

> The first is the State duty to protect against human rights abuses by third parties, including business enterprises, through appropriate policies, regulation, and adjudication. The second is the corporate responsibility to respect human rights, which means that business enterprises should act with due diligence to avoid infringing on the rights of others and to address adverse impacts with which they are involved. The third is the need for greater access by victims to effective remedy, both judicial and non-judicial.[34]

Seeking to "operationalize" this framework, Ruggie developed the follow-up Guiding Principles, which the Human Rights Council adopted in March 2011.[35] Covering both public and private actors, these principles make clear that both groups of actors have shared responsibilities in advancing human rights protection. Although the principles stop short of imposing on private actors the obligation to *protect* human rights, Principle 11 states in no uncertain terms that "[b]usiness enterprises should respect human rights." Principle 13 further identifies two sets of private human rights responsibilities:

> (a) Avoid causing or contributing to adverse human rights impacts through their own activities, and address such impacts when they occur;
> (b) Seek to prevent or mitigate adverse human rights impacts that are directly linked to their operations, products or services by their business relationships, even if they have not contributed to those impacts.

---

33 Human Rights Council, "Guiding Principles on Business and Human Rights: Implementing the United Nations 'Protect, Respect and Remedy' Framework" (2011) A/HRC/17/31 ("Guiding Principles").

34 Human Rights Council, "Report of the Special Representative of the Secretary-General on the Issue of Human Rights and Transnational Corporations and Other Business Enterprises" para 6, (2011) A/HRC/17/31.

35 Carlos López, "The 'Ruggie Process': From Legal Obligations to Corporate Social Responsibility?" in Surya Deva and David Bilchitz (eds), *Human Rights Obligations of Business: Beyond the Corporate Responsibility to Respect?* (Cambridge: Cambridge University Press 2013) 70–1.

4.3 Biobanks Formed Out of Public–Private Partnerships

Out of the three types of biobanks, analysis of the human rights obligations of those biobanks that are formed out of PPPs is the most challenging. Such analysis is challenging because these partnerships are hybrid entities that have "status as instruments of the public interest, yet bodies that actively engage private actors."[36] As a result, their human rights responsibilities do not align well with those traditionally assumed by either public or private actors.

Even more difficult, the human rights issues involving hybrid entities have been largely underresearched. The only document of which I am aware that has explored the human rights obligations involving PPPs in the intellectual property area is the Human Rights Guidelines for Pharmaceutical Companies in Relation to Access to Medicines.[37] These guidelines were developed in August 2008 by Paul Hunt in his capacity as the Special Rapporteur on the Right of Everyone to the Enjoyment of the Highest Attainable Standard of Physical and Mental Health. Although Guidelines 42 to 45 contain some rare provisions on PPPs, they do not fully delineate the partnerships' human rights responsibilities. Instead, the focus is on the human rights responsibilities of pharmaceutical companies in their role as the partnerships' public sector partners.

To address this lacuna, I suggested in an earlier work that the human rights responsibilities of a PPP can be determined by assigning responsibilities to three distinct entities: (1) the public sector partner; (2) the private sector partner; and (3) the partnership itself.[38] Such assignment aims to ensure that each set of responsibilities can be closely analyzed and monitored to determine whether each entity has appropriately discharged its human rights obligations. Because no entity can displace the human rights obligations of the other two, such an approach will also make analyzing the different sets of human rights obligations more manageable.

Under my proposed arrangement, the public sector partner will have the same obligations as any other public actor, even when the PPP which it forms with a private actor will have much lighter obligations. In the context of bio-

[36] Chris Skelcher, "Governing Partnerships" in Graeme A. Hodge, Carsten Greve, and Anthony E. Boardman (eds), *International Handbook on Public–Private Partnerships* (Cheltenham: Edward Elgar Publishing 2010) 292.

[37] Special Rapporteur on the Right of Everyone to the Enjoyment of the Highest Attainable Standard of Physical and Mental Health, "Human Rights Guidelines for Pharmaceutical Companies in Relation to Access to Medicines" (2008) A/63/263.

[38] Peter K. Yu, "Intellectual Property, Human Rights and Public-Private Partnerships" in Margaret Chon, Pedro Roffe, and Ahmed Abdel-Latif (eds), *The Cambridge Handbook of Public–Private Partnerships, Intellectual Property Governance, and Sustainable Development* (Cambridge: Cambridge University Press 2018).

banks, this public sector partner will assume obligations concerning the respect for privacy and autonomy, informed consent and the ability to withdraw such consent, the right to information and not to receive information, and protection against discrimination and stigmatization based on genetically related information derived from the biological materials collected by biobanks. These obligations concern issues already explored in the first section of this chapter.

In addition to these obligations, the public sector partner may have additional obligations imputed to it by virtue of its participation in a PPP, such as when it "fail[s] to exercise due diligence with respect to the [harmful] acts of [this partnership]."[39] It may also be held responsible for the partnership's human right breaches if it "ha[s] known, or should have known, that there was a real and immediate risk of [such breaches], and failed to take appropriate measures which might be expected to avoid this risk."[40]

Thus far, international organizations, regional bodies, and national courts have imputed human rights responsibilities to States based on factors such as functionality, organizational, or governance structure; management or financial control; dependency on the State; and public interest needs.[41] Although the scope and length of this chapter do not allow for a detailed exploration of the different ways of imputing human rights responsibilities to public actors,[42] the key takeaway of this truncated discussion is that the public sector partner cannot avoid its human rights obligations by outsourcing its operation to a PPP.

A similar analysis applies to the private sector partner. Just like any private actor, this private sector partner will assume the human rights obligations outlined in the Guiding Principles or other relevant human rights documents. For example, Principle 15 of the Guiding Principles calls on this partner to make a policy commitment to meet its human rights responsibilities, introduce a "human rights due-diligence process to identify, prevent, mitigate and account for how [it] address[es its] impacts on human rights," and institute "[p]rocesses to enable the remediation of any adverse human rights impacts [it] cause[s] or to which it contribute[s]." Principles 16 to 24 further delineate the specific human rights principles governing the operation of the private sector partner, in areas such as policy commitment, human rights due diligence, remediation, and issues of context.

Finally, as an independent entity, the PPP itself will have a distinct set of human rights responsibilities that are somewhat different from those assumed

---

39 Lisa Clarke, *Public–Private Partnerships and Responsibility under International Law: A Global Health Perspective* (Milton Park: Routledge 2014) 103.

40 *Ibid* at 119.

41 Christopher Bovis, *Public–Private Partnerships in the European Union* (Milton Park: Routledge 2014) 36–41.

42 Clarke, *supra* note 39, at 102–69.

by its constituent partners—whether public or private. These responsibilities will exist irrespective of the latter's obligations. Because the Guiding Principles do not provide clear guidelines on the human rights responsibilities of a PPP, this section offers some suggestions on what these responsibilities may entail and how the partnership can appropriately discharge them.

Like a private actor, the partnership will have to make a policy commitment to meet its human rights responsibilities, introduce a "human rights due-diligence process to identify, prevent, mitigate and account for how [it] address[es its] impacts on human rights," and institute "[p]rocesses to enable the remediation of any adverse human rights impacts [it] cause[s] or to which [it] contributes."[43] As part of human rights due diligence, the partnership "should identify and assess any actual or potential adverse human rights impacts with which [it] may be involved either through [its] own activities or as a result of [its] business relationships."[44] It should also be "prepared to communicate … externally [how it is to address its human rights impacts], particularly when concerns are raised by or on behalf of affected stakeholders."[45]

While the comparison with a private actor provides some baseline expectations, the partnership is a hybrid entity with dual characteristics as both a public actor and a private actor. Its human rights responsibilities therefore include those assumed by not only the former but also the latter. Although the partnership does not have the same regulatory power or the same ability to develop adjudicatory, grievance, or remedial mechanisms as States or other public actors, it can induce its public and private sector partners to better respect or protect human rights. For instance, similar to a public actor, the partnership can "[p]rovide effective guidance to [its constituent partners] on how to respect human rights throughout [its] operations."[46] It can also encourage its partners "to communicate how [it] address[es its] human rights impacts."[47] Such communication "can range from informal engagement with affected stakeholders to formal public reporting."[48]

As stated in the Guiding Principles, the partnership "should exercise adequate oversight in order to meet [the] international human rights obligations [of the public sector partner] when [the latter] contract[s] with, or legislate[s] for, business enterprises to provide services that may impact upon the enjoyment of human rights."[49] As the explanatory commentary on Principle 5 further

---

43 "Guiding Principles," *supra* note 33, Principle 15.

44 *Ibid*, Principle 18.

45 *Ibid*, Principle 21.

46 *Ibid*, Principle 3(c).

47 *Ibid*, Principle 3(d).

48 *Ibid*, Principle 3, explanatory commentary.

49 *Ibid*, Principle 5.

declares: "the relevant service contracts … should clarify the [public sector partner]'s expectations that these enterprises respect human rights." The partnership "should [also] ensure that [it] can effectively oversee the enterprises' activities, including through the provision of adequate independent monitoring and accountability mechanisms."[50]

In addition, drawing on Principle 8 of the Guiding Principles, the partnership should ensure that its public and private sector partners be "aware of and observe the [public sector partner]'s human rights obligations when fulfilling their respective mandates, including by providing them with relevant information, training and support." As noted in the explanatory commentary on this particular principle, attention should be paid to both vertical and horizontal coherence. While the former covers "the necessary policies, laws and processes to implement … international human rights law obligations," the latter concerns the ability of internal departments, agencies, or other sub-units "to be informed of and act in a manner compatible with [these] obligations."

Finally, we should not forget that some partnerships, due to their missions, designs, or structures, will take on the same level of human rights obligations that States or other public actors assume. Such heightened responsibilities can be attributed to the close nexus between these partnerships and the participating public actors, as well as the way the partnerships have been governed, financed, or supervised. Just as international organizations, regional bodies, and national courts may impute human rights responsibilities to States by virtue of their participation in PPPs, these bodies may also apply to these partnerships more stringent human rights standards that traditionally apply to public actors.

In sum, the wide variations in missions, designs, and structures may cause different types of PPPs to assume different levels of human rights responsibilities. Nevertheless, because of their hybrid nature and dual characteristics, virtually all biobanks formed out of PPPs will have to shoulder greater responsibilities than private biobanks, lest public biobanks have perverse incentives to avoid human rights obligations by forming PPPs. The degree of similarity between the responsibilities of a particular biobank formed out of a PPP and those of a public actor will largely depend on how closely the former resembles a public biobank and what nexus it has with a public actor, or multiple public actors.

---

50 *Ibid*, Principle 5, explanatory commentary.

5 CONCLUSION

Biobanks have provided many important benefits. Yet they have also raised many serious human rights questions. These questions range from the human rights issues implicated in the collection, processing, use, or storage of the biological materials collected by biobanks, to the human rights interests involved in the scientific productions utilizing the collected materials, to the human rights obligations of public biobanks, private biobanks, and biobanks formed out of PPPs. By providing this brief survey of the various human rights issues involved in the area of biobanking, it is my hope that this chapter will provide some useful groundwork for developing a deeper understanding of the many complex human rights issues involved. It is also my hope that the discussion of these issues will generate greater interest and attention from the medical community, biobank operators and funders, policymakers, regulators, commentators, and the mass media.

6. *You told me, right?* Free and informed consent in European patent law

Åsa Hellstadius and Jens Schovsbo[1]

1 INTRODUCTION

According to the European and EU (we use these expressions interchangeably) patent rules, inventions which involve (isolated) human biological material (HBM) are patentable if the general patentability conditions have been satisfied. A special rule provides, however, that for applications for such patents, the person from whose body the material is taken must have had the opportunity of expressing free and informed consent (FIC) in accordance with national law. In the following, we first locate the FIC requirement in European patent law and describe the (very limited) practice which has developed. Next, we turn to health law and fundamental rights law to better understand the role and function of FIC as a legal concept. Then we bring these insights back to patent law and use them to reconsider the present legal situation. We end with a discussion and conclusions on the relation between the requirement of FIC and patent law. The scope of analysis covers EU law, European patent law, and national law in Denmark and Sweden.

[1] This chapter is based on Åsa Hellstadius' presentation at Global Genes, Local Concerns: A Symposium on Legal, Ethical, and Scientific Challenges in International Biobanking, Copenhagen, March 16, 2017. Thanks to associate professor Janne Rothmar Herrmann Center for Advanced Studies in Biomedical Innovation Law (CeBIL), associate senior lecturer Ana Nordberg Lund University, and patent expert Patrick Andersson at the Swedish Patent and Registration Office for comments on an earlier draft.

2 FIC IN PATENT LAW

2.1 Introduction

Recital 26 of the Biotech Directive contains the following provision:[2]

> Whereas if an invention is based on biological material of human origin or if it uses such material, where a patent application is filed, the person from whose body the material is taken must have had an opportunity of expressing free and informed consent thereto, in accordance with national law.[3]

The provision raises a number of difficult questions: When is an invention "based" on material of human origin? At what point in time should the FIC have been obtained, and how specific should it be?[4] In this chapter, we merely note these complications as an inherent quality of the rule and do not attempt to provide any answers. Instead, we focus on the more basic question: *What are the legal effects of noncompliance with Recital 26?* Should our analyses lead us to conclude that noncompliance may affect the patentability of inventions based on HBM, the high complexity of the provision clearly suggests that its concepts and actual meaning may have a high practical impact.

Determining the legal effects of noncompliance with Recital 26 is inherently complicated. The first complication arises from the unclear legal status of the provision as a constitutional part of the Recitals of the Biotech Directive. It is

---

[2] Council Directive 1998/44/EC of July 6, 1998 on the legal protection of biotechnological inventions [1998] OJ L213, 13–21.

[3] Recital 26 should be understood in connection with Recital 16, which lays down the respect for fundamental principles of dignity and integrity of the person: "Whereas patent law must be applied so as to respect the fundamental principles safeguarding the dignity and integrity of the person" and Recital 43, which recognizes the international and constitutional foundations of respect for fundamental rights in the Directive: "Whereas pursuant to Article F(2) of the Treaty on European Union, the Union is to respect fundamental rights, as guaranteed by the European Convention for the Protection of Human Rights and Fundamental Freedoms signed in Rome on 4 November 1950 and as they result from the constitutional traditions common to the Member States, as general principles of Community law."

[4] By way of example, Deryck Beyleveld and Roger Brownsword, in *Human Dignity in Bioethics and Biolaw* (Oxford University Press 2001) 203, argue that "the most robust, literal, and indeed unambiguous interpretation of Recital 26" implies that FIC must be given "not only to the taking of the tissue *but also to the use of the tissue in research work that might lead to an application for a patent*" (emphasis in original). Detailed rules are found in Council Directive 2004/23/EC of March 31, 2004 on setting standards of quality and safety for the donation, procurement, testing, processing, preservation, storage and distribution of human tissues and cells [2004] OJ L102/48, 48–58, Article 13.

a general principle of EU law that "[w]hilst a recital in the preamble to a regulation may cast light on the interpretation to be given to a legal rule, it cannot in itself constitute such a rule."[5] Sometimes, however, this starting point is derogated. In order to sharpen the discussion we accept for the purposes of what follows the proposition by *Beyleveld*, *Brownsword*, and *Llewelyn* that this particular Recital should be considered as being prescriptive and thus as "legally binding."[6] Unlike these authors, we do not, however, infer from this in itself that noncompliance necessarily implies that EU Member States should be able to refuse patent applications if a FIC requirement has been violated or to find such patents to be invalid for that reason. In other words, we think that determining the legal effects of noncompliance is a separate issue from deciding on the binding character of the provision. In the following, we will first outline patent law's baseline for assessing patentability, and then zoom in on how FIC has been dealt with in national patent law by the Court of Justice of the European Union (CJEU) and the European Patent Office (EPO).

2.2 FIC in Patent Law

Recital 26 and the FIC requirement should be seen in the light of two basic principles of European patent law. The *first* basic principle is found in Article 3

[5] Case C-215/88 *Casa Fleischhandels-GmbH v Bundesanstalt für landwirtschaftliche Marktordnung* [1989] ECLI:EU:C:1989:273, para 31. Recitals are always subordinate to the operative provisions to which they relate. They have no operative effect on their own but a clear recital can limit the nature or the scope of an ambiguous operative provision. See Case C-136/04 *Deutsches Milch-Kontor GmbH v Hauptzollamt Hamburg-Jonas* [2005] ECR I-10095, para 32 and Case C-345/13 *Karen Millen Fashions Ltd v Dunnes Stores and Dunnes Stores (Limerick) Ltd* [2014] ECLI:EU:C:2014:2013, para 31. In the absence of an operative provision on FIC in the Directive, it is difficult to argue for such a construction from a legal–technical point of view. National implementation in Sweden and Denmark, which has made FIC compliance a matter of regulatory legislation other than patent law, is probably fulfilling the harmonization required by terms of the Directive.

[6] Deryck Beyleveld, Roger Brownsword, and Margaret Llewelyn, "The Morality Clauses of the Directive on the Legal Protection of Biotechnological Inventions: Conflict, Compromise and the Patent Community" in Richard Goldberg and Julian Lonbay (eds), *Pharmaceutical Medicine, Biotechnology and European Law* (Cambridge University Press 2001) 157, 175. Similarly (and pre the Judgment of the CJEU in Case C-377/98 *Netherlands v Parliament and Council* [2001] ECR I-07079), at 202–3 (describing the "European view" that insists upon there being FIC as a precondition for taking human material and as linking this obligation directly to Article 6(1) and thereby "inviting to opposition to patents where [FIC] requirements have not been properly satisfied"). In support of this view these authors refer to the Council of Europe Convention on Human Rights and Biomedicine (1997) Article 22 and the UNESCO Universal Declaration on the Human Genome and Human Rights (1997).

of the Biotech Directive, and is that inventions which satisfy the basic patentability criteria (that is, novelty, inventive step, and industrial applicability) shall be considered *patentable* even if they concern a product consisting of or containing biological material. Article 5 makes it clear that this principle extends to HBM, provided, however, that such material has been isolated from the human body or otherwise produced by means of a technical process, and (in relevant cases) that an industrial application of a sequence or a partial sequence of a gene has been disclosed in the patent application. In this way, EU law is in full conformity with the international baseline establishing the scope of patent law found in TRIPS Article 27(1),[7] that Member States shall provide for patent protection within "all fields of technology" for inventions which satisfy the basic patentability criteria.

The *second* basic principle limits the first one, and allows for a derogation from patentability for some types of inventions. Thus Article 6 of the Biotech Directive and Article 53(a) of the European Patent Convention (EPC) ("the morality exclusion") provides that inventions shall be considered *unpatentable* where their commercial exploitation would be contrary to *ordre public* or morality. Exploitation shall not be deemed to be so contrary merely because it is prohibited by law or regulation. According to the practice in the EPO (and in European national patent law generally) the application of these limitations is severely restricted, although the scope of the limitation, especially the specific "human embryo exclusion" in Article 6.2(c) of the Directive and Rule 28(c) EPC, has been somewhat broadened in the latest decisions concerning human embryonic stem cells.[8]

---

[7] Agreement on Trade-Related Aspects of Intellectual Property Rights (WTO) 1995.

[8] See G 2/06 (Use of Embryos/WARF) [2008] OJ 2009, 306 and T 522/04 (Stem cells/CALIFORNIA) [2009]. This broadening of scope of application is also evidenced by the practice of the CJEU. See Cases C-34/10 *Oliver Brüstle v Greenpeace eV* [2011] ECLI:EU:C:2011:669 and C-364/13 *International Stem Cell Corporation v Comptroller General of Patents, Designs and Trade Marks* [2014] ECLI:EU:C:2014: 2451, 7–8. It should be noted, however, that the broadening relates to the interpretation of the specific rule in regards to "uses of human embryos for industrial or commercial purposes" in Article 6.2(c) of the Directive and Rule 28(c) EPC, and we advocate caution in extending such a reading also to the general morality exclusion, not least in the absence of confirmation by case law. It can also be discussed whether this broadening of Article 6.2(c) and its counterpart in Rule 28(c) EPC are in conformity with TRIPS Article 27(2), according to which "Members may exclude from patentability inventions, the prevention within their territory of the commercial exploitation of which is necessary to protect *ordre public* or morality, including to protect human, animal or plant life or health or to avoid serious prejudice to the environment, provided that such exclusion is not made merely because the exploitation is prohibited by their law." See Åsa Hellstadius, *A Quest for Clarity: Reconstructing Standards for the Patent Law*

The basic principles of the Biotech Directive are mirrored in the EPC through the Implementing Regulations.[9] Although no express rule in the EPC corresponds to the contents of Recital 26 as such, the Directive should (in relevant cases) be used as a "supplementary means of interpretation" to the Convention.[10] According to the EPO Guidelines, the Recitals to the Directive are in particular to be taken into account.[11] As will be explained in Section 4.1, patent offices are obliged to assess on their own motion whether patent applications violate the morality exception and to reject those that do (*ex ante*). Patents which are issued but later found to violate the exclusion may be declared invalid, for example by national courts (*ex post*).

Following this schemata, inventions which fulfill the patentability criteria must be accepted for patenting *unless* the exclusion regarding *ordre public* or morality applies.[12]

2.3 Recital 26 in National Patent Law

By way of example, Recital 26 has been implemented in Denmark by the following provision in the Patent Order:

> If an invention relates to or makes use of a biological material of human origin, it *shall appear* from the patent application whether the person from whom the biological material originates has given his consent to the filing of the application. The information about consent *shall not affect the examination and other processing of the patent application or the validity of the rights conferred by the granted patent* (emphasis added).[13]

Morality Exclusion (Stockholm University 2015) 403ff and Ana Nordberg and Timo Minssen, "A 'Ray of Hope' for European Stem Cell Patents or 'Out of the Smog into the Fog'? An Analysis of Recent European Case Law and How It Compares to the US" (2016) International Review of Intellectual Property and Competition Law 138, 162ff.

[9] See Article 53(a) EPC and Chapter V, Rules 26–34 of the Implementing Regulations of the EPC.

[10] Rule 26(1) of the Implementing Regulations of the EPC.

[11] Guidelines for examination in the European Patent Office (November 2018), G-II, 5.2. The express mentioning of recitals as a means of interpretation of the EPC may point in the direction of treating FIC as an important factor in the patentability of human biological material. On the other hand it would be difficult to argue that such a requirement should function at the same level as the express criteria in the Articles and Rules of EPC.

[12] None of the other exclusions in, for example, Article 53 EPC are applicable to the issue of FIC.

[13] Section 3(6) of the Danish Order No 25 of January 18, 2013 on Patents and Supplementary Protection Certificates.

The Danish rule treats Recital 26 as "legally binding" but makes it clear that noncompliance is not sanctioned in patent law.[14] Instead, violating this rule could constitute a violation of the Danish Criminal Act Section 162 on incorrect statements to a public authority.[15] In other words: this legal setup accepts the legally binding nature of Recital 26 but has decoupled the consequences of nonperformance from the patent system. Whereas this system arguably is in conformity with the obligation arising from a literal interpretation of Recital 26, it has obviously reduced the effect of the Recital, since the scope of the provisions in the Criminal Act is very limited and has (to our knowledge) never been relied upon in practice.

Contrary to the Danish case, Recital 26 has not been introduced as such in Swedish patent legislation.[16] Instead, the rules on integrity, and specifically consent, in the Act on Ethical Review of Research and the Act on Biobanks apply to the use of HBM for research leading up to an invention.[17] These Acts have no connection with patent law.[18] Noncompliance with consent rules and standards will have no effect on the patentability of the resulting invention, since the requirement of consent is not part of the patentability criteria in the Swedish Patents Act.[19] This means that a person whose material is used in an invention has no procedural means to stop its patenting or to get a granted patent invalidated. Although according to preparatory works the examiners at the Swedish Patent and Registration Office must pay specific attention to patent applications concerning biotechnological inventions which may raise ethical concerns, they are not required (by terms of the law) or even allowed to refuse grant in situations where noncompliance with FIC rules may be an issue.[20]

---

14 The wording "shall appear" (skal … fremgå) implies a formal criterion. This is, however, not the case due to the express statement that the examination or processing of the patent application is not affected by the information.

15 According to the Criminal Act Section 162 anyone who makes an incorrect statement before any public authority or for the information of such authority concerning matters on which he is bound to give evidence shall be liable to a fine or to imprisonment for any term not exceeding four months. The provision in Recital 27 on the identification of the origin of biological material is sanctioned in the same way.

16 Patentlag (1967:837).

17 Lag (2003:460) om etikprövning av forskning som avser människor, Lag (2002:297) om biobanker i hälso- och sjukvården m.m.

18 Violation of informed consent rules can lead to fines, prison sentences and/or damages. See Chapter 6 of the Act on Biobanks and Section 38 of the Act on Ethical Review of Research.

19 Swedish Government Committee Report 2008:20, Patentskydd för biotekniska uppfinningar, 266–7; 300ff.

20 Conclusion by the Swedish Government Committee on Patent Protection for Biotechnological Inventions, see Government Report 2008:20, 290.

2.4 Recital 26 in EU Law

The CJEU has on one occasion dealt with the principle of FIC in the context of patent law, in a case concerning the validity of the (then) newly enacted Biotech Directive.[21] The applicants (Kingdom of Netherlands etc) submitted an action for the annulment of the Directive based upon six pleas. The fifth plea concerned the breach of the fundamental right to respect for human dignity, and consisted of the argument that the absence of a provision requiring verification of the consent of the donor or recipient of products obtained by biotechnological means undermined the right to self-determination, thereby undermining the principle of human integrity. In addressing this issue, the CJEU made the following remarks:

> 78. The second part of the plea concerns *the right to human integrity*, in so far as it encompasses, in the context of medicine and biology, *the free and informed consent of the donor and recipient.*
> 79. Reliance on *this fundamental right* is, however, clearly misplaced as against a directive which concerns only the grant of patents and whose scope does not therefore extend to activities before and after that grant, whether they involve research or the use of the patented products. (emphasis added)[22]

As can be seen, the CJEU at the same time identified FIC as a fundamental right and made it clear that the requirement concerns "only the grant of patents."[23] This statement arguably opens space for two apparently conflicting lines of argumentation for understanding the role and effect of Recital 26, viz what we call the "regulatory rule line of argumentation" and the "fundamental rights line of argument."

The *regulatory rule* line of argumentation suggests that noncompliance does not have the potential to affect the validity of patents and should not be considered as being a part of the (substantive) patentability requirements (via the morality exclusion). This is the model preferred by the Danish and Swedish legislators, previously discussed. Arguably it is also the option preferred by the CJEU, judging from the content of the statement in para 79. This position implies a restrictive approach to Recital 26. It also seems to imply that in the absence of a procedure for making FIC an actual substantive patentability criterion, the principle has no bearing upon the examination and grant of patents. This does not necessarily imply that the FIC obligation is not seen as being "legally binding," but the legal effects are dealt with *outside* of substan-

21 Judgment of the CJEU in case C-377/98, *supra* note 6.
22 *Ibid.*
23 Hellstadius, *supra* note 8, at 367.

tive patent law. This can be interpreted in part as a result of the Directive's restricted scope of application to only the actual granting of patents and not to activities before or after the process of examination and grant.

The *fundamental rights* argument, on the contrary, seems to be boosting the role and importance of Recital 26 as a fundamental right. This line of reasoning begs an understanding of the provision as part of Article 6. As was explained earlier, such a reading suggests that noncompliance may affect patentability/validity.

In this way, the judgment did little to demystify the provision apart from confirming the Advocate General's remark regarding its lack of clarity.

The CJEU has not addressed the issue of FIC directly in other judgments. It is interesting to note, however, that the judgment in *Brüstle* (on human embryonic stem cells) seemingly challenges the Court's remark that the Directive "concerns only the grant of patents and whose scope does not therefore extend to activities before and after that grant." In this case, the Court interpreted the exclusion in Article 6 of the Directive as allowing a broad range of matter to be assessed, including such activities that are not included in the patent application (and therefore not directly included within the scope of the Directive) and activities that preceded the invention both in material aspects (such as derivation of material) as well as in time.[24] It is, therefore, by no means clear whether the CJEU would consider FIC compliance to have an actual bearing on the patentability assessment.[25]

2.5 Recital 26 in EPO Case Law

As a general rule, Article 53(a) EPC should only be applied in rare and extreme cases, where the commercial exploitation of an invention threatens

---

[24] Case C-34/10, *supra* note 8, para 49: "an invention must be regarded as unpatentable, even if the claims of the patent do not concern the use of human embryos [that is, excluded subject matter], where the *implementation of the invention* requires the destruction of human embryos" (emphasis added).

[25] The European Group on Ethics in Science and New Technologies (EGE) has expressed that an invention based on HBM ("elements of human origin") having been retrieved without respecting the principle of consent will not fulfill the ethical requirements. See Opinion No 8, Ethical aspects of patenting inventions involving elements of human origin, September 25, 1996, para 2.4. Opinion issued by the EGE predecessor GAEIB (The Group of Advisers on the Ethical Implications of Biotechnology). See Geertrui Van Overwalle, "Gene Patents and Human Rights" in Paul L.C. Torremans (ed) *Intellectual Property Law and Human Rights* (3rd edn, Kluwer Law International 2015) 871, 902ff.

ordre public or morality.[26] The EPO has had severe difficulties with the interpretation and application of this exclusion, resulting in legal uncertainty.[27] Nevertheless, it seems to be generally accepted that ethical principles which lie behind the concept of morality are anchored at the highest level in the legal hierarchy of norms.[28] Whether or not violation of national law also constitutes a threat to *ordre public* or morality is an issue which has not yet been decided.[29]

Furthermore, the EPO uses two different concepts to measure the gravity of morality/*ordre public* violations. In some decisions, the EPO has stated that inventions whose exploitation does not conform with the conventionally accepted standards of conduct are to be excluded from patentability, denoting acceptance (or rather nonacceptance) of the public at large as the decisive criterion for the application of the morality exclusion.[30] In other decisions, the standard employed has been one of abhorrence, meaning that only those inventions that would be universally regarded as outrageous should be excluded—indicating a more severe standard for application. These variations in application have led to a situation where nonpatentability by terms of the morality exclusion is assessed on a case-by-case basis.

The "moral guardian function" of Article 53(a) has led the EPO to consider FIC (non)compliance in a few decisions concerning donated HBM. HBM have formed the basis for research leading up to a patent application, or even formed part of the invention as such. The case law of EPO's Boards of Appeal has developed along the two separate lanes of argumentation identified previously. Thus, the *fundamental rights* line of argumentation would seem to have been expressed in the *Relaxin* case. In this, the Opposition Division found that the (lawful and informed) consent given by the women to the act of isolation of the tissue in question (which formed the basis of the invention), actually countered the objections raised on moral grounds to the patentability of the invention.[31]

---

26 See, for example, Guidelines for the Examination in the European Patent Office (November 2018), G-II, *supra* note 11, at 4.1.

27 The interpretation and scope of Article 53(a) is widely discussed in legal doctrine. For an exposé of the different arguments, see Hellstadius, *supra* note 8, at 202ff.

28 See, for example, Joseph Straus, "Patenting Human Genes in Europe—Past Developments and Prospects for the Future" (1995) IIC 26(6) 920, 932; Ulrich Schatz, "Patents and Morality" in Sigrid Sterckx (ed) *Biotechnology, Patents and Morality* (Edward Elgar 1997) 159, 161; Rainer Moufang, "The Concept of 'Ordre Public' and Morality in Patent Law" in Geertrui Van Overwalle (ed) *Patent Law, Ethics and Biotechnology* (Bruylant 1998) 65, 71; Deryck Beyleveld and Roger Brownsword, *Mice, Morality and Patents* (Common Law Institute of Intellectual Property 1993) 56ff.

29 Hellstadius, *supra* note 8, at 202ff.

30 See, for example, T 356/93 (Plant cells/PLANT GENETIC SYSTEMS) of 21.2.1995, OJ 1995, 545, para 6.

31 V 8/94 (Howard Florey/Relaxin), OJ EPO 1995, para 6.1.

In a similar vein, evidence of compliance with legal FIC requirements led to a positive decision on patentability in the *Leland Stanford* decision, where the Opposition Division, reflecting the view in *Relaxin*, held that the use of HBM for research "is widely accepted provided that consent was given, which there is no reason to doubt in the present case."[32] These decisions seem to adhere to the *fundamental rights* line of reasoning, where the respect for FIC is expressed as a fundamental right and patentability is confirmed where such consent has been lawfully collected.

In contrast to *Relaxin* and *Leland Stanford*, the *regulatory rule* line of reasoning was followed in the *Breast and Ovarian Cancer* decision.[33] The Board stated that since the legislator had not provided for a procedure of verifying the informed consent in the framework of the grant of biotechnological patents under the EPC, patent law was not the appropriate framework for the imposition and monitoring of such a requirement.[34]

Thus, the few decisions of the EPO Boards of Appeal regarding FIC are based on conflicting lines of reasoning. Consequently, no clear guidance as to how the FIC requirement should be handled has emerged. Considering the small number of decisions, however, conclusions must be drawn with caution. On the one hand, the EPO seems perfectly content to use FIC compliance as a positive argument for patentability, which could be interpreted as a *fundamental rights* line of reasoning. On the other, since this particular line of reasoning has only ever been expressed where the result is positive (that is, where FIC compliance supports patentability), one cannot be certain that this line would be used and/or hold up in a situation of noncompliance with FIC. As evidenced by the reasoning in *Breast and Ovarian Cancer*, the EPO seems reluctant to accept that noncompliance with FIC might affect patent validity (and grant), thereby choosing the *regulatory rule* line of reasoning.

In sum, none of the cases we have studied here provide for clear and uniform answers to the question of what legal effects might be brought about by noncompliance with FIC. Following the decisions of the CJEU and EPO, it is hard to discern whether noncompliance affects patentability (and validity). In the following, we turn to health law and fundamental rights for insights into the function and role of FIC.

---

[32] Opposition Division decision of 16.08.2001 (LELAND STANFORD/Modified Animal), Reasons for the Decision, para 8(50).

[33] T 1213/05 (Breast and ovarian cancer/UNIVERSITY OF UTAH) of 27.9.2007.

[34] *Ibid*, Reasons for the Decision, para 47.

3 FIC IN HEALTH LAW AND FUNDAMENTAL RIGHTS LAW

3.1 Fundamental Rights

The FIC requirement is one of the fundamental principles of research ethics, which originated in the aftermath of the atrocities of World War II, especially the abuses during the Holocaust. During the Nuremberg Trials, ten standards were formulated which physicians must follow when carrying out experiments on human subjects. According to the first standard of the Nuremberg Code, the voluntary consent of the human subject is absolutely essential. This standard, and other fundamental ethical principles,[35] were later formalized and codified by international and national legislation.[36] Today, the FIC principle is expressly codified in a number of international fundamental rights conventions and charters.

Physical and mental integrity have long been recognized as enjoying protection as part of the right to private life guaranteed by Article 8 of the European Convention on Human Rights (ECHR).[37] The Council of Europe International Biomedicine Convention (Oviedo Convention), signed by most of the European States, sets out the fundamental principles applicable in

---

35 For biomedical research involving humans, the development of bioethics as a necessary discipline of medical and biological research has had a profound impact on integration of ethical principles in legislation. But a number of principles remain just principles, and their influence on legislation and decisionmaking is often a question of national institutional means, such as the setting up of research ethics committees or establishing guidelines for procedures, such as providing standardized informed consent forms. National legislation is furthermore often general and vague, international declarations may lack a specific legal status on the national level, and, more importantly, a large section of principles is regulated by so-called soft law in the forms of codes of conduct by professional organizations—interpreted on a case-by-case basis of authorities and agencies.

36 The Nuremberg Code 1947. See also the United Nations (UN) Universal Declaration of Human Rights 1948, the European Convention for the Protection of Human Rights and Fundamental Freedoms 1950, the UN International Covenant on Civil and Political Rights 1966, the UN International Covenant on Economic, Social and Cultural Rights 1976, the Council of Europe Convention for the Protection of Human Rights and Dignity of the Human Being with regard to the Application of Biology and Medicine: Convention of Human Rights and Biomedicine (Oviedo) 1997 and the World Medical Association Helsinki Declaration on Ethical Principles for Medical Research Involving Human Subjects 1964.

37 Sabine Michalowski, "Article 3—Right to the Integrity of the Person" in Steve Peers and others (eds), *The EU Charter of Fundamental Rights: A Commentary* (Hart 2014) 39, 43.

day-to-day medicine as well as those applicable to new technologies in human biology and medicine.[38] It draws on the principles established by the ECHR in the field of biology and medicine. Thus, several of the ECHR provisions overlap, in terms of application, with provisions in the Oviedo Convention in the field of bioethics and human rights.

Article 5 of the Convention expressly rules:

> An intervention in the health field may only be carried out after the person concerned has given *free and informed consent* to it.
>
> This person shall beforehand be given appropriate information as to the purpose and nature of the intervention as well as on its consequences and risks.
>
> The person concerned may freely withdraw consent at any time. (emphasis added)[39]

The principle of free and informed consent is further reinforced by Article 22 (Disposal of a removed part of the human body):

> When in the course of an intervention any part of a human body is removed, it may be stored and used for a purpose other than that for which it was removed, only if this is done in conformity with appropriate information and consent procedures.

Thus, informed consent is necessary not only for interventions on persons, but also for so-called secondary uses for HBM removed from the human body. The principle in Article 22 is confirmed by Additional Protocol No 195,[40] on biomedical research.[41] On the basis of protection of human dignity and respect for human integrity, persons participating in research projects shall be specifically informed not only of the immediate purpose(s) of the research, but also of "any foreseen potential further uses, *including commercial uses*, of the research results, data or biological materials".[42]

---

[38] Council of Europe Convention for the Protection of Human Rights and Dignity of the Human Being with regard to the Application of Biology and Medicine: Convention of Human Rights and Biomedicine (Oviedo 1997).

[39] See Part II of the Oviedo Convention. Articles 6–9 set out specific rules on consent concerning, for example, emergency situations.

[40] The content of the Oviedo Convention may be supplemented by various Additional Protocols, such as on human cloning (1998), organ transplantation (2002), and genetic testing for health purposes (2008).

[41] Council of Europe Treaty Series—No 195, "Additional Protocol to the Convention on Human Rights and Biomedicine" (Strasbourg, January 25, 2005) https://rm.coe.int/168008371a accessed 27 October 2017, concerning Biomedical Research.

[42] Articles 1 and 13.2.vii of Protocol No 195. See also Van Overwalle, *supra* note 25, at 903ff.

The Oviedo Convention is binding on states parties to the agreement, and the principles contained therein need to be integrated by effective measures in national law.[43] In Denmark the Convention has been in force since December 1, 1999, with some reservations. The articles of the Convention are not, however, incorporated into Danish law, as is the case with the European Convention on Human Rights.[44] Sweden has not ratified the Oviedo Convention yet, but it was signed on April 4, 1997.[45] Ratification status varies among the European states.[46] The lack of ratifications will perhaps not have a profound effect on the interpretation of FIC requirement, due to the influence of Article 8 ECHR and the linking of the two instruments by the case law of the European Court of Human Rights (ECtHR).

The ECtHR may give Advisory Opinions on legal questions concerning the interpretation of the Oviedo Convention.[47] This competence reinforces the strong link between the two instruments.[48] But the Court may only examine violations of the ECHR.[49] Even so, due to the overlap between the two Conventions there are judgments concerning Articles of the ECHR that provide foundation for the application of provisions of the Oviedo Convention, which naturally affects states' obligations in this respect.[50] This would also have the effect that so far as FIC issues fall under Article 8 of the ECHR, even the states that have not (yet) signed or ratified the Oviedo Convention would need to give effect to such interpretation under the ECHR.[51]

The right to integrity in the form of FIC also has a limited basis in Article 7 of the UN International Covenant on Civil and Political Rights, where it is

---

43 Article 1 of the Oviedo Convention.

44 See Janne Rothmar Herrmann, *Retsbeskyttelsen af fostre og ubefrugtede æg—Om håndteringen af retlige hybrider* (DJØF Publishing 2008) 26.

45 The main reason for the lack of ratification is that Swedish legislation does not fulfil the requirements regarding the protection of persons not able to consent (Article 6 of the Oviedo Convention). Sweden has therefore not been able to adhere to the Additional Protocols either.

46 See Council of Europe, "Chart of signatures and ratifications of Treaty 164" www.coe.int/en/web/conventions/full-list/-/conventions/treaty/164/signatures?p_auth=Egk0fvQZ accessed October 11, 2017.

47 Article 29 of the Oviedo Convention. Individuals do not, however, have the right to bring proceedings before the Court. See Paragraph 165 of the Explanatory Report to the European Convention on Human Rights and Biomedicine.

48 Francesco Seatzu and Simona Fanni, "The Experience of the European Court of Human Rights with the European Convention on Human Rights and Biomedicine" (2015) 31(81) Utrecht Journal of International and European Law 112–13.

49 See Article 32 of the ECHR.

50 See, for example, *Costa and Pavan v Italy* App no 54270/10 (ECHR, 28 August 2012) and *Evans v the United Kingdom* App no 6339/05 (ECHR, April 10, 2007).

51 Seatzu and Fanni, *supra* note 48, at 113.

stated that "no one shall be subjected without his free consent to medical or scientific experimentation."

The content of Article 5 of the Oviedo Convention is further encoded in Article 3 of the Charter of Fundamental Rights of the European Union,[52] *Right to the integrity of the person*, which states the following:

1. Everyone has the right to respect for his or her physical and mental integrity.
2. In the fields of medicine and biology, the following must be respected in particular:
   a. *the free and informed consent of the person concerned, according to the procedures laid down by law*,
   b. the prohibition of eugenic practices, in particular those aiming at the selection of persons,
   c. the prohibition on making the human body and its parts as such a source of financial gain,
   d. the prohibition of the reproductive cloning of human beings. (emphasis added)

As pointed out previously, the CJEU in the *Netherlands* case recognized "the right to human integrity" as a fundamental right which should be secured by the Court in its review of the compatibility of EU legislation. The fundamental right to dignity and integrity pervades all areas of EU law and policy.[53] The importance of FIC has also been underlined in a number of EU Directives, viz the Human Tissue Directive,[54] the Organ Safety Directive,[55] and the Clinical Trials Directive.[56] All of these Directives leave it to the EU states to decide on the details on FIC, and the reference to "according to the procedures laid down by law" in Article 3.2(a) also points to the necessity of rules in national legislation for the (procedural) functioning of the FIC requirement. But according to one commentator, the express recognition of FIC in the Charter's Article 3.2 "confirms its fundamental importance, and sets a minimum standard for the applicable consent laws of the Member States".[57]

---

52 OJ C326, 391–407.

53 Michalowski, *supra* note 37, at 41.

54 Council Directive 2004/23/EC, *supra* note 4.

55 Directive 2010/45/EU of July 7, 2010 on standards of quality and safety of human organs intended for transplantation [2010] OJ L207/14, Article 14.

56 Directive 2001/20/EC of April 4, 2001 on the approximation of the laws, regulations and administrative provisions of the Member States relating to the implementation of good clinical practice in the conduct of clinical trials on medicinal products for human use [2001] OJ L121/34, Article 3(2).

57 Michalowski, *supra* note 37, at 47.

It follows from Article 51(1) of the Charter that it is binding not only on the EU but also on the EU countries when they are implementing EU law. A provision in the Charter cannot, however, be applied by itself vis-à-vis private parties; it needs the existence of a least one relevant rule of EU law other than a Charter provision. The Charter is also intended to have a *ripple effect*. As a source of primary EU law, the Charter—or fundamental rights in general—has been invoked before the CJEU in many cases across the whole spectrum of EU law, although the Court's attitude toward the scope and its effects for individuals may vary.[58] In order for the Charter to be *exemplary*, the Commission has also adopted a "Strategy for the effective implementation of the Charter of Fundamental Rights by the European Union" as part of an ambitious fundamental rights policy.[59]

Against this background, there is no doubt that FIC is an important fundamental right and, as such, an expression of the principle of respect for human integrity. National authorities have a duty to adhere to the minimum standards of FIC set out in the international instruments to which the national states are parties. The express confirmation of FIC as part of primary EU law by Article 3 in the EU Charter reinforces the impact and status this requirement should enjoy in national law.

The right to integrity finds its basis in the concept of human dignity, which is not only a fundamental right in itself but is also recognized as the basis for all fundamental rights.[60] Dignity refers to the intrinsic worth of all humans and is the foundation of inalienable human rights, including autonomy, which protects the capacity to make one's own decisions and necessitates respect for the decisionmaking capacity of others.[61]

The link between integrity and dignity is especially visible with regard to the principle of FIC, since any conduct interfering with a person's sense of integrity or self-worth—for instance, medical intervention without consent—may

---

58 Allan Rosas, "Five Years of Charter Case Law: Some Observations" in Stephen Weatherill, Ulf Bernitz, and Sybe de Vries (eds), *The EU Charter of Fundamental Rights As a Binding Instrument: Five Years Old and Growing* (Bloomsbury 2015) 11, 13.

59 Commission, "Strategy for the effective implementation of the Charter of Fundamental Rights by the European Union" (Communication) COM (2010) 573 final. As part of this endeavor the Commission publishes annual reports on the application of the Charter: see http://ec.europa.eu/justice/fundamental-rights/charter/application/index_en.htm accessed May 10, 2018 (none of these contain references to the issue of consent).

60 See the Preamble of the 1948 Universal Declaration of Human Rights.

61 Roger Brownsword, "Bioethics Today, Bioethics Tomorrow: Stem Cell Research and the Dignitarian Alliance" (2012) 17(1) Notre Dame Journal of Law, Ethics & Public Policy 15, 42.

be regarded as an attack on that person's dignity.[62] Thus, FIC is fundamentally an expression of respect for human dignity.

Because individuals have dignity they have also rights, such as the right to make autonomous decisions. And these rights can only rarely be overridden, as exemplified by Article 5 of the Helsinki Declaration: "In medical research on human subjects, considerations related to the well-being of the human subject should take precedence over the interests of science and society."[63] According to this statement, respect for human dignity must thus rule out utilitarian approaches to the issue of FIC, or at least lead to the conclusion that preserving human dignity in the form of FIC always overrules any other interest at stake.

The intrinsic value of human dignity and its impact on regulation of biotechnological research, especially regarding humans, cannot be overestimated. Despite the importance of the dignity concept as a founding fundamental right, its contents remain elusive. In bioethics, different conceptions of dignity have led to tension and debates about its scope and use, especially in relation to regulation of biotechnology—something that has allegedly led to different meanings of the concept being applied in legal instruments, both European and international.[64] The contested nature of human dignity in the regulation of biotechnological research may result in a lack of clarity about the effects of human dignity concerns, and difficulties for researchers who are anxious to achieve regulatory compliance in regards to the FIC requirement.[65] This may occur in particular when ethics committees decide on FIC in situations where they need to base decisions on general references to human dignity. Debates about, for example, human dignity implications for biotech patenting are not a new phenomenon.

In addition, an important question should be raised when discussing FIC in the context of research on HBM. The status of FIC as a fundamental right and a prerequisite for research on human beings is inviolable. But what if the focus is not primarily on the human body as a human being, but rather on the future use of the HBMs which are separated, collected, and stored in the form of biological samples? Even though Article 22 of the Oviedo Convention, along

---

62 Michalowski, *supra* note 37, at 47.

63 Timothy Caulfield and Roger Brownsword, "Human Dignity: A Guide to Policy-Making in the Biotech Era?" (2006) Nature Reviews Genetics 7, 73.

64 The prevailing approaches are human dignity as empowerment and human dignity as constraint. See, for example, Aurora Plomer, "Human Dignity, Human Rights, and Article 6(1) of the EU Directive of Biotechnological Inventions" in Aurora Plomer and Paul Torremans (eds), *Embryonic Stem Cell Patents, European Law and Ethics* (Oxford University Press 2009) 209ff; Caulfield and Brownsword, *supra* note 63, at 72.

65 *Ibid* at 75.

with its Protocol 195, explicitly expresses that such secondary uses must also be subject to free and informed consent procedures, the legal status of FIC in relation to such separated HBM is much more heterogeneous, not least on the national level. It may be noted that the actual scope and content of the principles of respect for human dignity and integrity have been considered from different standpoints by ethicists in these situations.[66]

3.2 Health Law

The use of residual tissue removed in the course of diagnosis or therapy in research without FIC used to be the norm. This has changed, however, and now it is widely accepted that donors must be asked for permission to store and use their samples.[67] The term "FIC" covers a very heterogeneous group of different models ranging from very specific consents, through broad consents and dynamic ones, to inferred consents. The choice of model may depend on the aim and context of the taking of material, research and procedures, and, more importantly, the balancing of interests of the stakeholders involved.

The principle of voluntary donation of HBM in accordance with FIC procedures is permeating the legal framework, regardless of the purpose of the donation, such as research or health care (for example, transplantation). Questions about the type of consent needed for collection, storage, and future research use of donated human material—such as blood or cell or tissue material—for yet to be specified research are among the most challenging for legislators and research communities. The boundaries between purposes may blur, and changing circumstances may lead to samples collected and stored for one purpose subsequently being used for secondary purposes.[68] For instance, samples stored for clinical or research purposes may end up as important parts of commercial products.

The emergence of large-scale population projects, involving a great number of participants, poses a challenge for FIC procedures, especially when material is used for new projects. Often, however, the information forming the basis of the donor's consent is limited to the immediate purpose of the procedure, which is often treatment and/or research, and does not include, for example, future

---

66 See for example Brownsword, *supra* note 61, at 42; Plomer, *supra* note 64, at 207.

67 See Cameron Stewart and others, "The Problems of Biobanking and the Law of Gifts" in Imogen Goold and others (eds), *Persons, Parts and Property: How Should We Regulate Human Tissue in the 21st Century* (Bloomsbury 2014) 27.

68 Carlo Petrini, "Ethical and Legal Considerations Regarding the Ownership and Commercial Use of Human Biological Material and Their Derivates" (2012) 3 Journal of Blood Medicine 87.

commercial purposes.[69] Designing a FIC procedure foreseeing all possible future uses is of course difficult, and raises many ethical and legal questions.

In Sweden and Denmark, ethics review committees play an important role in safeguarding the FIC compliance for research on human material. The donation and use of human biological material in Sweden is covered by several different legal acts: the Genetic Integrity Act,[70] the Act concerning the Ethical Review of Research Involving Humans,[71] the Biobanks in Medical Care Act,[72] the Quality and Safety Standards in the Processing of Human Tissue and Cells Act,[73] the Transplantation Act,[74] and so on. All of these acts regulate, to different extents, the necessary consent procedures, the scope of which may vary. For instance, a donor of tissue samples for collection and storage in a biobank must be informed of the intention and purpose (or purposes) of the biobank in question and must give his or her consent.[75] Documentation in the patient's medical record is mandatory.[76] New FIC by the donor is, however, required for further use of samples—except if the new purpose is research or clinical trials, in which case the board for ethics review may allow exceptions.[77] For research involving studies on biological material from identifiable dead or living humans, consent must be freely given, explicit, specified, and documented (normally in writing).[78]

Failure to comply with the FIC rules will normally result in sanctions in the form of imprisonment, fines, or perhaps damages (the latter in relation to the individual), depending on the type of legal act that regulates the specific situation.[79]

In Denmark, all research projects involving human beings or any kind of human tissue, cells, and so on need permission from a regional research ethics committee.[80] FIC is required as a general rule, and if a cell or tissue sample is

---

69 Cf *ibid* at 93.

70 Lag (2006:351) om genetisk integritet m.m.

71 Lag (2003:460) om etikprövning av forskning som avser människor.

72 Lag (2002:297) om biobanker i hälso- och sjukvården m.m.

73 Lag (2008:286) om kvalitets- och säkerhetsnormer vid hantering av mänskliga vävnader och celler.

74 Lag (1995:831) om transplantation m.m.

75 Chapter 3, Section 2 of the Biobanks Act.

76 *Ibid*, Section 7.

77 *Ibid*, Section 5.

78 Sections 13 and 16–22 of the Ethical Review Act.

79 See, for example, Chapter 6 of the Biobanks Act and Section 38 of the Ethical Review Act.

80 See Chapter 3 of the Act on Biomedical Research (593/2011) (Lov om videnskabsetisk behandling af sundhedsvidenskabelige forskningsprojekter) and Chapter 4ff. of the Health Act (1188/2016) (Sundhedsloven).

taken for research purposes and stored in a biobank, written FIC is the norm. Noncompliance with FIC rules may result in imprisonment or fines.[81]

An important difference between Denmark and Sweden is the regulation of FIC for secondary use of biological samples, that is, where previous FIC has been collected for a particular purpose but where the sample in question is valuable for a secondary use, for example for research. In Danish law, presumed consent is the current standard, unless a person has registered in the Use of Tissue Registry (opt-out system), meaning they do not wish their cells or tissue to be used for anything other than the treatment and diagnosis of themselves.[82] In Sweden and the other Nordic countries, explicit rather than presumed FIC is the departure point for secondary use of samples, meaning that new FIC must be collected, but a number of exceptions from this main rule exist.[83]

To sum up, although FIC is an important prerequisite in national health law in both Denmark and Sweden for research on HBM, and as such is subject to detailed regulation, some situations are unregulated and usually subject to the assessment of ethics committees. Few cases of noncompliance with FIC rules exist and decisionmaking by relevant authorities is rather the norm, making the requirement subject to soft law.

Thus, there is little doubt that the requirement for FIC enjoys a status as a fundamental human right, as an expression of the principles of human integrity and dignity. As such, it cannot be compromised or overridden by other concerns. The study reveals that in our example countries, Denmark and Sweden, FIC is regulated in important legal acts but there are also lacunas, not least concerning use of HBM in the many situations where assessment of FIC necessity or relevance is a matter for research ethics committees. The heterogeneity of the FIC concept in law, coupled with the unclear scope of application of human dignity concerns for the use of HBM separated from the human body, may in practice make regulatory compliance difficult.

4 FINDINGS AND DISCUSSION

4.1 Patent-Procedural Aspects of FIC Compliance

The question we set out to answer is: *what are the legal effects of noncompliance with Recital 26 in European patent law?* From our study we can

---

81 See Chapter 9 of the Act on Biomedical Research.

82 Section 32 of the Health Act.

83 See, for example, Chapter 4 of the Swedish Biobanks Act; Salla Silvola, "Biobank Regulation in Finland and the Nordic Countries" in Elisabeth Rynning and Mette Hartlev (eds), *Nordic Health Law in a European Context: Welfare State Perspectives on Patients' Rights and Biomedicine* (Martinus Nijhoff 2011) 277, 281ff.

conclude that at present there is no clear answer to that question. We have found that both the EPO and the CJEU act vaguely and inconsistently with regard to the scope and application of the respective morality exclusions. Both have, on occasion, argued in line with the *regulatory rule* line of reasoning and excluded questions of FIC compliance from the scope of assessment of patentability.[84] On the other hand, evidence of decisions which include of FIC considerations in the patentability assessment also exists.[85] In this way, two competing narratives are present at the same time: according to the *regulatory rule* approach, noncompliance with FIC does not affect the validity or the question of patentability; on the other hand, the *fundamental rights* approach means noncompliance has the potential to cause a patent denial or to invalidate patents.

We find that the *fundamental rights* approach offers the most compelling narrative. The FIC requirement is an expression of one of the fundamental rights, namely that of human dignity, more specifically in the form of human integrity. As an overriding norm in terms of legal status compared with patent law rules, it should not be possible to refrain from assessing effects of noncompliance, especially since additional support is provided for including respect for FIC rules when such are relevant for an invention, in terms of an actual provision (albeit a Recital) in the Biotech Directive. The *fundamental rights* argument clearly acknowledges this connection between patent law and *fundamental rights* norms and suggests a (necessary) solution in the inclusion of Recital 26 (FIC compliance) in the application of the morality exclusion in Article 6 of the Biotech Directive.

In light of the status of FIC, conceptualizing it in patent law as a purely procedural issue which does not have the potential to affect the validity of patents becomes problematic. As mentioned, the mere presence of Recital 26, a norm which clearly makes the fulfillment of FIC principles a matter for the patenting process, cannot be overlooked. From a legal–technical point of view, Recital 26 has no corresponding substantive provision in the Directive, and as such has no binding legal force on its own. But even though Recital 26 would not be able to stand on its own as a substantive rule, its presence has another, broader effect. Recital 26 articulates the status of FIC as a fundamental legal norm *intra* patent law. Such status cannot be negated by resorting to arguments on procedure, more specifically the lack of procedural rules. And by the terms

---

84 See case C-377/98, *supra* note 6 and decision T 1213/05, *supra* note 48.

85 See case C-34/10, *supra* note 8 (although FIC was not part of the decision, the CJEU expanded the scope of application of Article 6(2) of the Biotech Directive to activities before filing of the patent application), V 8/94 (Howard Florey/Relaxin), OJ EPO 1995, and Opposition Division decision of 16.08.2001 (LELAND STANFORD/ Modified Animal).

of Article 6 of the Directive, supported by Recital 26, FIC noncompliance not only can but should be raised as an argument of breach of the fundamental right of integrity under the umbrella of the general morality exclusion.

If one accepts that FIC compliance falls within the scope of the morality exclusion, the absence of a specific procedure for its verification does not exempt patent offices from considering this aspect as part of their examination. Divisions and Boards of the EPO (at least) thus have a responsibility to identify and assess relevant issues on their own motion because of the *ex officio* principle found in in Article 114 EPC. According to this, "[i]n proceedings before it, the European Patent Office shall examine the facts of its own motion; it shall not be restricted in this examination to the facts, evidence and arguments provided by the parties and the relief sought." The content of this principle is explained in the Guidelines:

> [O]nce proceedings have been initiated, if there is reason to believe, e.g. from personal knowledge or from observations presented by third parties, that there are facts and evidence not yet considered in the proceedings which in whole or in part prejudice the granting or maintenance of the European patent, such facts and evidence must be included in those examined by the EPO of its own motion pursuant to Art. 114(1).[86]

Arguably, this obligation extends to issues which might trigger the morality exclusion. Thus, accepting FIC as part of the scope of the morality exclusion would necessitate an investigation of evidence on its merits.

With regard to questions related to morality, the first stage examination is naturally conducted on the basis of the patent application, while the appeal review may widen the scope of relevant issues.[87] This follows from the double nature of the patenting procedure, where the administrative setting at the examining stages takes on a judicial nature on appeal.[88] The civil nature of the EPO *inter partes* proceedings puts the responsibility on the initiating party to

---

[86] Guidelines for examination in the European Patent Office (November 2018), *supra* note 11, Part C, Chapter VIII-7. See also Catarina Holtz, "Due Process for Industrial Property: European Patenting under Human Rights Control" (Doctoral thesis, Handelshögskolan i Stockholm 2003), 272, on the effects of Article 114 for opposition proceedings and the reasoning by Hellstadius, *supra* note 8, 287–91.

[87] See for example G 10/93 (*Ex parte* examination) and Holtz, *ibid*, at 74. See also Hellstadius, *supra* note 8, at 143ff.

[88] Holtz, *supra* note 86, at 63, with further references to Jacobsson et al, *Patentlagstiftningen—en kommentar* (P A Norstedt & Söners förlag 1980) 236, Benkard et al, *Patentgesetz Gebrauchmustergesetz*, 9. C.H. Auflage, *Beck'sche Verlagsbuchhandlung* (München 1993) 49; G. Paterson, *The European Patent System: The Law and Practice of the European Patent Convention*, 2nd edn (Sweet & Maxwell 2001) 165 at 4–70; Bernitz et al, *Immaterialrätt och otillbörlig konkurrens*, 7th edn

present legal facts in order to fulfill the relevant prerequisite. In the matter of morality and *ordre public* this means that the party must invoke the relevant norm corresponding to the legal prerequisite and the issue relative to a breach of that norm.[89] This must be substantiated by the facts relevant to the issue and it is ultimately the authority that decides on the norm in question. The EPO appeals procedure is also sufficiently broad to provide the Boards of Appeal with full powers according to "the language of the law" to decide on whether a patent shall be awarded or not, even where the scope of the issues go beyond the requests and facts presented by the parties.[90] Judging from the existing bulk of EPO decisions regarding the morality exclusion, it is also usually on appeal that the relevant facts, evidence, and arguments take form—often in the form of objections from opponents. Nevertheless, it is important to remember that the *inter partes* nature of patent examination does not relieve the patent authority from its task of identifying and assessing morality issues already at the examination stage.[91] And to the extent that noncompliance with FIC is a matter for the morality exclusion, then it also falls to the examiner to properly identify and assess such issues.[92]

The Swedish Patents and Trademark Registration Office (PRV) has in a few cases checked patent applications on its own motion for compliance with rules of approval of research by Ethical Review Councils.[93] This is as close to a procedural review as a patent authority can come without express legal procedures for FIC verification.

Against this background, we recommend that courts and patent authorities accept the fundamental rights approach as the starting point for dealing with FIC. We believe that there are simply no compelling arguments not to at least open a space for assessment of FIC compliance within the patent process for HBM inventions. This recommendation has implications for the way patent authorities deal with the issue of FIC, which we will address next.

(Jure Förlag 2001), 106, and M. Koktvedgaard and M. Levin, *Lärobok i immaterialrätt*, 7th edn (Norstedts Juridik 2002) 195 f.

89 See Holtz, *supra* note 86, at 81–2.

90 *Ibid* at 272–3. See also Article 117 EPC on evidence.

91 *Supra* note 9.

92 In this context, see the possibility of hearing of experts in Rule 117 EPC (Decision on taking of evidence) as well as the possibility of addition of a legally qualified examiner during the initial examination phase (Guidelines for Examination in the European Patent Office (November 2018), for example D-V, C-VIII). This latter possibility is explicitly recognized in relation to Article 53(a) EPC, see Guidelines G-II, 4.1 and C-VIII, 7. See also Holtz, *supra* note 86, at 74ff.

93 Åsa Hellstadius, "A Comparative Analysis of the National Implementation of the Directive's Morality Clause" in Aurora Plomer and Paul Torremans, *supra* note 64, at 117, 132.

4.2 FIC Implications for Patentability: A Balancing of Interests

Having accepted the high-ranking nature of the FIC obligation, one would expect patent authorities to vigorously monitor its observance. However, given the heterogeneity of the FIC concept, such monitoring would be exceedingly complicated for patent authorities. For this reason, we think that the acceptance of this recommendation presupposes that clear instructions are provided for patent authorities in deciding whether or not the FIC criteria have been met. This is an inherently complicated matter. Would, for instance, blanket consent suffice, or should consent have been given explicitly for patenting, or only where this is a foreseeable possibility?[94] Here we acknowledge that patent authorities would have to consider relevant provisions in regulatory law and perhaps seek additional guidance from other regulatory bodies, for instance research ethics committees.[95]

While we believe it is important for the patent system to acknowledge the importance of the individual donor as expressed in the principle of FIC, we also find that the overall balancing of interests, including those of patentees and general societal interests, in an efficient and well-functioning patent system must be maintained. These considerations seem to suggest that one should not automatically consider noncompliance with FIC as having the ability to affect patent validity. Giving individual donors a broad right to oppose patenting or to successfully challenge the validity of patents based on noncompliance with FIC procedures would seem to be highly problematic.

This stance is partly based upon the functioning of the morality exclusion as a last resort and absolute safeguard, with a narrow scope and adhering to a high threshold for application, excluding only such inventions that are truly immoral (unacceptable, abhorrent) from patentability. Furthermore, the morality exclusion is only triggered if the invention is regarded as reprehensible by *society in general*.[96] In order for a FIC violation to pass this high threshold, the mere reliance of a patent on HBM which has been obtained in violation of a FIC requirement would be insufficient. The requirement that societal values

---

94 Petrini, *supra* note 68, at 93.

95 It should be noted that the EPO has refrained from tying the interpretation of the morality exclusion to existing legal rules, neither national nor international. This is understandable from a historic perspective, with the forming of a European Patent Organisation comprising many national states and a view to a foreseen unitary patent. Conversely, national authorities have one set of legislation and one state's cultural norms to consider, which evidently could still be a difficult task but nonetheless more manageable. See Hellstadius, *supra* note 8, at 408ff.

96 T 866/01 (Euthanasia compositions/MICHIGAN STATE UNIV.) (11 May 2005) para 6.12, and Hellstadius, *supra* note 8 at 208.

have to be compromised would seem to imply that harm suffered by individuals would not pass muster.[97] For donors to be given such a possibility a general moral wrong resulting from the patenting itself of a particular piece of HBM, which is somehow connected to one single person, must be established. We find it very hard to see that mere noncompliance with an FIC rule would normally amount to such a wrong.

On the other hand, the legal status of FIC and the absence of a procedure for verification do not release the patent authorities from the requirement to evaluate the effects of proof of absence of FIC in a patent application based on HBM, if such an objection is raised under Article 53(a) EPC (or becomes a matter of interpretation by the CJEU with regard to Article 6 of the Biotech Directive). Consequently, in those instances where noncompliance of FIC could be said to constitute an infringement not just of the individual donor but of *ordre public* or morality, that is, of the general ethical norms of high value, such a noncompliance should—and must—be allowed to threaten the validity of patents. Given the absolutely exceptional nature of these situations, it is difficult to predict exactly which types of noncompliance would be interpreted as an infringement of the morality exclusion and result in nonpatentability. Granted, collecting and using samples from donors without complying with FIC requirements at all, or deceiving donors by giving wrong or misleading information about the purpose of the samples (for instance, promising use for clinical treatment but instead using samples for commercial purposes)—perhaps also targeting donors in vulnerable situations—would definitely amount to a violation of most regulatory rules of FIC, and the added moral wrong resulting from such acts could, and perhaps should, invoke the application of the morality exclusion. It should be added, however, that even in such instances the ability to rely on the morality exception should probably be limited to instances in which there is a direct, causal link between the donation of the sample (relying on the defective FIC) and the invention.

For these reasons, we think it should be accepted as a matter of principle that nonconformity with FIC requirements may constitute a factor which, through

---

[97] The question of whether individuals can assert property rights to donated HBM is accompanied by continuous debates. See for example Dianne Nicol et al, "Impressions on the Body, Property and Research" and Simon Douglas, "Property Rights in Human Biological Material," both in Goold et al (eds) *Persons, Parts and Property*, 9–24 and 89–108. Of interest is also the possibility for individuals to enter into contractual solutions in regards to their donation of HBM. See also Morten E. Juul Nielsen, Nana C. Halmsted Kongsholm, and Jens Schovsbo, "Property and Human Genetic Information," Journal of Community Genetics, https://doi.org/10.1007/s12687-018-0366-4 (April 30 2018) (arguing that donors do not have any inherent "rights" to control "their" genetic information also regarding patenting).

the morality exclusion, has the potential to invalidate concrete patents, but also that only certain types of noncompliance should have the potential to do so. We also think that such instances should be rare.

Should patent authorities accept our recommendations, they would need to intensify the scrutiny of patent applications to make sure that the monitoring of FIC compliance is possible for patent applications regarding HBM. We would also recommend that steps are undertaken to clarify what kind(s) of consent would normally be acceptable, which should normally amount to legal requirements in relevant regulatory acts. Noncompliance with such FIC should arguably be legally sanctioned, but should normally be channeled outside of patent law—for example, in the Danish case, toward the Criminal Code. Unlike the situation under the *regulatory rule* line of reasoning previously identified, our approach would open a space for challenges to the validity of patents based on the rules on morality, but only in rare instances.

5 FINAL REMARKS

Based on our analyses, we make the following recommendations.

First, patent authorities should accept that compliance with FIC requirements falls under the scope of the morality exclusion, should be subject to monitoring *ex officio*, and has the potential to lead to nonpatentability of HBM inventions or invalidity of already granted patents. *Second*, the potential for FIC noncompliance to lead to the nongranting or invalidation of patents should be limited to rare and extreme circumstances where there is a direct and casual link between the donation of the sample (relying on the defective FIC) and the invention, and where the wrong suffered by the donor amounts to a violation of general ethical norms.

We believe such a system would be best suited to reflect the overall balancing of interests involved in the patenting of HBM, which involves both individual and governance interests and general societal interests in innovation. At the same time, such an approach would recognize the overriding value and importance of the FIC principle as a fundamental ethical principle founded in respect for human dignity and integrity.

7. Dynamic Consent and biobanking: a means of fostering sustainability?

Jane Kaye and Megan Prictor

1 INTRODUCTION

Biobanks are rich repositories of biological materials (such as DNA) and other health and demographic data, often collected over a long period, that can be used for a variety of research purposes to improve the health of individuals and populations. It is important that the value of biobanks is maximized, but at this point in time there are a number of challenges to achieving this.[1] There is continued debate over the most appropriate mode of gaining consent from people who contribute tissue samples and data to biobanks, which will uphold high ethical standards and enable autonomous decisionmaking. As in other fields, there are changing legal and regulatory frameworks that can have significant implications for biobank management. There is also increasing concern as to whether biobanks are achieving maximum usage and the nature of longer-term sustainability plans for maintaining these repositories. In this chapter, we will outline some of the risks facing biobanks, using examples drawn from a range of international settings. We suggest that the concept of "Dynamic Consent," a digital platform for engaging research participants, has the capacity to ensure a more engaged and informed cohort of participants, which might in turn address many of the legal and sustainability challenges currently facing biobanks. We examine current uses of Dynamic Consent platforms in biobanking research in the UK, continental Europe, and the USA, and outline considerations for future application and evaluation of this tool to help enhance the relevance, ethical operation, sustainability, and interoperability of biobanks.

[1] European Commission and Directorate-General for Research and Innovation, *Biobanks for Europe: A Challenge for Governance* (EUR-OP 2012) 13.

2 THE CHALLENGES FACING BIOBANKS

Over the past 20 years there has been considerable investment in biobanks around the world to support population research, and national biobank hubs have been built through initiatives such as the BBMRI-ERIC within Europe.[2] There has also been sizable investment in bioinformatics and analytic platforms, but there has not always been a corresponding investment in digital technology to engage with participants. Some of the key challenges that are currently faced by biobanks, such as consent, dealing with complex regulatory environments, and long-term sustainability, could be improved by using digital technologies to engage with biobank participants.

2.1 Obtaining Consent

Consent is one of the ongoing issues that could be greatly improved using digital technologies. Consent is a foundation principle in the ethical conduct of research, and in international data protection law. The World Medical Association's Declaration of Helsinki (1964) requires that potential subjects of medical research "must be adequately informed of the aims, methods, sources of funding, any possible conflicts of interest, institutional affiliations of the researcher, the anticipated benefits and potential risks of the study and the discomfort it may entail, post-study provisions and any other relevant aspects of the study."[3] Consent mechanisms for research usually involve a one-time written consent—a method that is well established,[4] although expensive to administer and not always comprehensible to or well recalled by participants.[5]

---

[2] BBMRI-ERIC, "A Gateway for Health—Biobanking and BioMolecular Resources Research Infrastructure" www.bbmri-eric.eu/ accessed December 12, 2017. See also: Michaela T. Mayrhofer and others, "BBMRI-ERIC: The Novel Gateway to Biobanks. From Humans to Humans" (2016) 59 Bundesgesundheitsblatt—Gesundheitsforschung—Gesundheitsschutz 379; Eero Vuorio, "Networking Biobanks Throughout Europe: The Development of BBMRI-ERIC," *Biobanking of Human Biospecimens* (2017) <https://link.springer.com/chapter/10.1007/978-3-319-55120-3_8> accessed December 12, 2017.

[3] World Medical Association, "WMA Declaration of Helsinki—Ethical Principles for Medical Research Involving Human Subjects" www.wma.net/policies-post/wma-declaration-of-helsinki-ethical-principles-for-medical-research-involving-human-subjects/ accessed November 20, 2017.

[4] Anne Tassé, "Is Written Informed Consent Outdated?" (2017) 27 European Journal of Public Health 195, 195.

[5] D. Strech and others, "A Template for Broad Consent in Biobank Research. Results and Explanation of an Evidence and Consensus-Based Development Process" (2016) 59 European Journal of Medical Genetics 295.

The informed consent approach assumes that people are making a decision about their personal involvement in one study, which is limited in time and to geographic boundaries, and whose methods, risks, and benefits can be anticipated.

In the context of biobanking, when samples and data are collected their future research use may be unspecified, and perhaps unforeseen. It could be intended, or at least possible, that they are later shared "across organizational and national boundaries."[6] Whether biobanks are disease-specific or population-based, they often contain genetic data, meaning that consideration must be given to recontacting donors (for instance, if an incidental finding with an actionable health risk is identified) and to the involvement of or disclosure to other family members. Specific consent to each study, as conceived of under traditional norms and policies, is thus challenging in this setting. As an approach, it is a poor fit for many features of biobank activity, including the potentially very large number of participants, the large number of studies that cannot be foreseen at the time of participants' enrollment, and the long period during which contact and recontact may be needed. The cost and resource implications of seeking specific consent from each participant to each use of their biobank sample may be substantial and impractical.

As a result, research involving biobanks has often adopted a broad consent approach, intended to cover many or all potential future uses of research samples. Some researchers, prompted by awareness of the time and effort involved in obtaining approval by research ethics committees for projects involving biobank samples (and particularly research across jurisdictions), have pursued "even broader consent for possible future uses than perhaps is envisaged in the general discussion of broad consent in the literature … many felt it was the only solution for secondary research because going back for a new consent for a new research use meant going back through the cumbersome research ethics process again."[7] For biobanks there is widespread acceptance of broad consent, and it forms part of the World Medical Association (WMA) Declaration of Taipei on Ethical Considerations Regarding Health Databases and Biobanks, as well as the Organisation for Economic Co-Operation and Development (OECD) Guidelines on Human Biobanks and Genetic Research Databases.[8] Nonetheless, there is not yet a consensus that broad consent for

---

6 E.A. Whitley, N. Kanellopoulou, and J. Kaye, "Consent and Research Governance in Biobanks: Evidence from Focus Groups with Medical Researchers" (2012) 15 Public Health Genomics 232, 237.

7 *Ibid* at 240.

8 World Medical Association, "WMA Declaration of Taipei on Ethical Considerations Regarding Health Databases and Biobanks" para 12 www.wma.net/policies-post/wma-declaration-of-taipei-on-ethical-considerations-regarding-health

future unforeseen uses of samples and data is appropriate from an ethical or legal perspective,[9] and consent for biobanking remains a subject of debate for academic scholars.[10]

2.2 Legal Framework

Instead of being uniform and static data repositories, biobanks are highly differentiated, both by the nature of their collections and by their policy, regulatory, and legal contexts, which differ both by location and over time. Research using biobanks often involves international collaborations and the sharing of samples and data across jurisdictions, which may involve obtaining access to multiple biobanks. Such cross-border sharing of genetic data and biological samples has been encouraged as a way to promote scientific advancement and international cooperation, as outlined in article 18 of the UNESCO International Declaration on Human Genetic Data.[11] The navigation of disparate laws and policies that is necessary to do this is not straightforward, however, and creates impediments for researchers.

The legal and governance frameworks that apply to biobanks are "nationally fragmented"[12] and change over time. They are usually not specific to biobanks, but rather regulate data protection, privacy, or human research generally. Samples and data respectively may fall under different laws. In the UK, requirements relevant to consent in genomic research (as an example) have spanned the Clinical Trials Regulations, the Data Protection Act 1998, and the Human Tissue Act 2004, with its Codes of Practice.[13] Across Europe, the General Data Protection Regulation that came into effect in May 2018 has direct, binding application to member states in the European Union,[14]

-databases-and-biobanks/ accessed December 10, 2017; OECD, "OECD Guidelines on Human Biobanks and Genetic Research Databases" para 4.6 www.oecd.org/sti/biotech/44054609.pdf accessed December 12, 2017.

[9] Zubin Master and others, "Biobanks, Consent and Claims of Consensus" (2012) 9 Nature Methods 885.

[10] Ubaka Ogbogu and others, "Newspaper Coverage of Biobanks" (2014) 2 PeerJ.

[11] UNESCO, "International Declaration on Human Genetic Data" www.unesco.org/new/en/social-and-human-sciences/themes/bioethics/human-genetic-data/ accessed August 28, 2017.

[12] Jane Kaye and others, "Access Governance for Biobanks: The Case of the BioSHaRE-EU Cohorts" (2016) 14 Biopreservation and Biobanking 201, 202.

[13] Paula Boddington and others, "Consent Forms in Genomics: The Difference between Law and Practice" (2011) European Journal of Health Law 491, 495.

[14] The full title of this Regulation is: "Regulation (EU) 2016/679 of the European Parliament and of the Council of 27 April 2016 on the protection of natural persons with regard to the processing of personal data and on the free movement of such data, and repealing Directive 95/46/EC."

but its goal of providing a unified data protection regime across Europe is undermined by the opportunity for national derogations for research.[15] The Regulation also substantially tightens the requirements for consent to data collection and processing, bringing into question the use of past broad consent by biobank participants at the time of enrollment as a basis for future studies using their data.[16] To take another example, across African countries there is a widespread absence of regulation for genomic research and biobanking. This has the effect of pushing decisions about issues such as consent and access to individual biobank governing bodies and research ethics committees, rather than their being decided upon by lawmaking authorities.[17]

The complexity of legal and regulatory frameworks is evident in many jurisdictions. Australia provides an example of a federated legal structure, with laws at both Commonwealth and state levels affecting biobank activities.[18] The Privacy Act 1988 (Cth) regulates biobanks' management of health information, and is combined with state-based legislation on health records (for instance, the Health Records Act 2001 (Vic)) and human tissue.[19] There is no Commonwealth legislation on health and medical research; instead, the independent statutory agency, the National Health and Medical Research Council (NHMRC), issues guidelines on relevant matters, such as the National Statement on Ethical Conduct in Human Research (2007) (updated May 2015), which sets out the requirements for data use by custodians and researchers.[20]

---

[15] Information Commissioner's Office, "Exemptions" https://ico.org.uk/for-organisations/guide-to-the-general-data-protection-regulation-gdpr/exemptions/?q=derogations accessed April 20, 2018.

[16] Information Commissioner's Office, "Consultation: GDPR Consent Guidance" https://ico.org.uk/media/about-the-ico/consultations/2013551/draft-gdpr-consent-guidance-for-consultation-201703.pdf accessed December 9, 2017.

[17] Jantina de Vries and others, "Regulation of Genomic and Biobanking Research in Africa: A Content Analysis of Ethics Guidelines, Policies and Procedures from 22 African Countries" (2017) 18 BMC Medical Ethics 8.

[18] Jane Kaye and others, "Trends and Challenges in Biobanking," in Ian Freckelton and Kerry Petersen (eds), *Tensions and Traumas in Health Law* (Federation Press 2017).

[19] Human tissue legislation is in place in all Australian jurisdictions. See: Human Tissue Act 1983 (NSW); Transplantation and Anatomy Act 1979 (Qld); Transplantation and Anatomy Act 1983 (SA); Human Tissue Act 1985 (Tas); Human Tissue Act 1982 (Vic); Human Tissue and Transplant Act 1982 (WA); Transplantation and Anatomy Act 1978 (ACT); Human Tissue Transplant Act 1979 (NT).

[20] The National Health and Medical Research Council, the Australian Research Council, and the Australian Vice-Chancellors' Committee, "National Statement on Ethical Conduct in Human Research 2007 (Updated May 2015)" 3.2.3–3.2.8 www.nhmrc.gov.au/_files_nhmrc/publications/attachments/e72_national_statement_may_2015_150514_a.pdf accessed July 24, 2017.

The absence of targeted legislation for biobanking in Australia and elsewhere impedes interoperability across state and national boundaries because of the complexity of determining compliance requirements in different locations and for different types of research. It has been argued that the complexity, fragmentation, and mutability of biobank regulation incur substantial costs in time and effort for researchers and may have the effect of impeding largescale research endeavors.[21]

This diversity of legal and governance frameworks internationally also impacts on research participants who donate samples and data to biobanks. One effect is that different participants are likely to experience involvement in research quite differently depending on the jurisdiction and the nature of the research project.[22] For example, consent mechanisms are designed to fit pre-existing local requirements rather than being harmonized across biobank infrastructures in different settings and locations. Boddington and colleagues confirm that "there are no detailed national or international mandatory standards for the format of consent forms for research participants,"[23] and so the forms are an accrual of past practice at each location. This also means that the information that is provided to individuals is specific to individual biobanks and so may vary in terms of comprehensiveness and quality between biobanks. This fragmentation has the potential to result in a lack of uniform and equal protections for research participants as these are dependent upon the jurisdiction in which they are based, but also the integrity of local biobank managers and regional best practice.

3 BIOBANK USAGE AND SUSTAINABILITY

Biobanks have been the subject of substantial enthusiasm and investment since the turn of the century, and were described in *Time* magazine in 2009 as one of "10 ideas changing the world right now."[24] At the time of writing, the value of the global biobanking market is heading towards US$200 billion per annum.[25] National governments are continuing to spend large amounts of taxpayer money around the world to establish population biobanks, but also to support national infrastructure and specialist collections. To justify this expenditure,

---

21 Kaye and others, *supra* at note 12, 202.

22 Kaye and others, *supra* at note 18.

23 Boddington and others, *supra* at note 13, 492.

24 Judita Kinkorová, "Biobanks in the Era of Personalized Medicine: Objectives, Challenges, and Innovation: Overview" (2015) 7 EPMA Journal 4.

25 Jackson Highsmith, "Biobanking: Technologies and Global Markets" www.bccresearch.com/market-research/biotechnology/biobanking-technologies-markets-report-bio084b.html accessed November 21, 2017.

biobanks need to demonstrate that they are good value for money but also that they are also making a significant contribution to the benefit of the wider society. Biobanks need to be able to demonstrate that they are being used for a range of different research purposes and that they are relevant to researchers so that they can attract funding and investment. To achieve this, biobanks need to be nimble and responsive to changing circumstances so they can meet researchers' needs and remain relevant to changing research priorities.

However, there is growing concern around usage and sustainability, considering this substantial investment and biobanks' degree of accountability to the general public. Low utilization of biobanks[26]—so-called biohoarding[27]—may not only undermine sustainability but also call into question the fundamental purpose of biobanks. Researchers have expressed deep frustration with the difficulty of getting access to required tissue samples and of obtaining ethical approval, both of which involve lengthy delays and laborious paperwork.[28] Some commentators have called for a policy reorientation, suggesting that biobanks should be judged by "the knowledge generated by the facility and a guarantee of the delivery of health service improvements" rather than the number of specimens collected, in order to realize the potential of biobanks to enable new research discoveries in human disease and behavior rather than simply amassing collections for their own sake.[29]

For many biobanks, the goal is to obtain long-term sustainable funding rather than being reliant on project grants. This transition has been achieved by biobanks being embedded in translational research systems and becoming an intrinsic part of a healthcare system. At the same time, there has also been a considerable debate around whether to charge researchers for access, and on what basis—a calculation which largely depends upon individual biobank circumstances. Many biobanks have realized sustainability requires cost-efficient storage, management, and organization of samples and data. It also requires the ability to obtain new samples, but also different types of samples on a regular basis, and to link this with medical records and other related information. The ideal is that a biobank should be able to recontact participants easily and quickly, which depends upon good communication systems as well as an ongoing and committed relationship with participants, which could be achieved by using digital technologies.

---

26 Megan Scudellari, "Biobank Managers Bemoan Underuse of Collected Samples" (2013) 19 Nature Medicine 253.

27 Daniel Catchpoole, "'Biohoarding': Treasures Not Seen, Stories Not Told" (2016) 21 Journal of Health Services Research & Policy 140, 140.

28 Whitley, Kanellopoulou, and Kaye, *supra* at note 6, 236–7.

29 Catchpoole, *supra* at note 27, 141.

4 WHAT IS DYNAMIC CONSENT?

Dynamic Consent has been described both as an approach and as a specific tool to facilitate interaction between research participants and researchers. As an approach, it is conceptualized as a flexible, sustained online communication platform to enable the provision of information to participants, and their tailored specification and/or withdrawal of consent to participation in research.[30] It allows for people's ongoing engagement with research in a way that reflects their personal preferences. In the context of biobanking, a Dynamic Consent tool can enable potential participants to give broad consent or to specify in advance that their permission must be sought for each new use of their samples or data. They can revisit and change these specifications over time, including withdrawing their consent and being assured that this has taken effect.

Researchers, similarly, can use a Dynamic Consent platform to provide initial information to potential participants, invite those who already have contributed to a biobank to allow the use of their samples or data in new studies, and deliver reports back to participants on how and by whom these samples and data are being used. Use of the platform can extend beyond consent and communication to incorporate the collection of participants' self-reported health and lifestyle data.

Dynamic Consent has been operationalized in several research projects around the world. In South Tyrol, Italy, Dynamic Consent has been built into a prospective population-based biobank known as CHRIS (Cooperative Health Research in South Tyrol), which collects blood and urine samples and health and lifestyle data to illuminate the interaction between genetic, lifestyle, and environmental factors in the development of common diseases.[31] The project commenced in 2011, and by 2017 more than 10,000 people had participated.[32] The principle of recontacting participants for further data collection over time is fundamental to the CHRIS study, requiring long-term communication and trust between participants and researchers. To this end, a Dynamic Consent tool provides an indispensable mechanism both for initial consent decisions and for subsequent decisions and communication over several years.

---

30 Isabelle Budin-Ljøsne and others, 'Dynamic Consent: A Potential Solution to Some of the Challenges of Modern Biomedical Research" (2017) 18 BMC Medical Ethics 4.

31 "CHRIS Gesundheitsstudie—Vinschgau Südtirol—Eurac Research" https://de.chris.eurac.edu/ accessed June 5, 2018.

32 Cinzia Piciocchi and others, "Legal Issues in Governing Genetic Biobanks: The Italian Framework as a Case Study for the Implications for Citizen's Health through Public–Private Initiatives" (2017) Journal of Community Genetics 1.

After a video explaining the project is shown, electronic consent is given online, directly on the personal interactive consent webpage. The type of consent asked for is broad with regard to the aim of the study. At the same time, the consent is layered and provides dynamic options (changeable online over time) regarding data sharing (international, public data repositories), return of secondary/unexpected results (outlining the right to know or the right not to know), and the permission to use samples and data in case of death or if the subject loses legal capability. The data regarding access levels granted by each participant goes directly into the database and ensures that data can automatically be filtered for different purposes according to the participants' choices. The dynamic tool can also be used for recontact, collecting additional information, and reconsent, should this be necessary in the future.[33]

In the UK-based 'RUDY' study (Rare UK Diseases of bone, joints and blood vessels), conducted over five years, a Dynamic Consent tool has been used to give people nuanced control over their decision to participate in sub-studies of the overarching research, including the capacity to revisit and amend their choices over time.[34] Dynamic Consent options can be tailored in terms of researchers' access to people's health records, use of samples and data, and continuing contact.[35] This clinical research network has a biobank embedded within it and the Dynamic Consent options enable research participants to determine whether they wish their data and samples to be shared beyond the clinical research team.

Dynamic Consent has also been incorporated into a US-based platform called "PEER" (Platform for Engaging Everyone Responsibly), which is used by more than 45 organizations collecting data on more than 15,000 participants.[36] Additional functionality being built into that tool will facilitate data capture from electronic medical records, and provide information in multiple languages. Dynamic Consent is being considered for research involving people with myotonic dystrophy in Japan,[37] where scholars have also devoted

---

33 *Ibid.*

34 Harriet J.A. Teare and others, "The RUDY Study: Using Digital Technologies to Enable a Research Partnership" (2017) 25 European Journal of Human Genetics 816.

35 M.K. Javaid and others, "The RUDY Study Platform—A Novel Approach to Patient Driven Research in Rare Musculoskeletal Diseases" (2016) 11 Orphanet Journal of Rare Diseases 150.

36 Genetic Alliance, "Platform for Engaging Everyone Responsibly" www.geneticalliance.org/programs/biotrust/peer accessed November 9, 2017.

37 Victoria Coathup and others, "Using Digital Technologies to Engage with Medical Research: Views of Myotonic Dystrophy Patients in Japan" (2016) 17 BMC Medical Ethics 51.

attention to expanding the tool to incorporate family involvement.[38] It is also anticipated that the National Centre for Indigenous Genomics in Canberra, Australia may incorporate a Dynamic Consent platform for its repository of indigenous biospecimens, genomic data, and documents.[39] Researchers and citizens alike recognize a range of benefits in a Dynamic Consent approach compared with traditional forms of consent, in terms of enhanced two-way communication, the efficient exercise of privacy rights, and a richer data set through the possibility of additional sample and data collection. These benefits will be outlined in more detail in what follows, considering the ways in which Dynamic Consent might meet some of the challenges facing biobanks in terms of sustainability and interoperability.

5 THE BENEFITS OF DYNAMIC CONSENT

Implemented for use with biobanks, ranging from those at the small scale to others with a major, national, or cross-border scope, a Dynamic Consent approach to involving participants is likely to benefit researchers, participants, and the research endeavor itself.

5.1 Ethical and Legal Standards

As outlined earlier, the legal and regulatory context for biobanking internationally is marked by complexity, fragmentation, and change over time. The "broad consent" approach that is often adopted for biobanking may not be well understood or preferred by research participants and the wider community. Arguably, single-instance written consent forms completed at the time of initial donation to a biobank are a poor mechanism for researchers seeking to uphold high standards of ethical research conduct. Nor is it clear that they enable biobanks to meet the requirements of data protection laws, because consent is not being given for downstream use of sensitive information for different purposes.

A Dynamic Consent approach enables participants to choose not only their initial involvement but also the degree to which they wish to exercise choice in subsequent participation; further, it permits reconsideration of the consent decision, and a withdrawal action that has clear effect. These features of Dynamic Consent mean that it meets the highest international ethical and

---

38 Jusaku Minari and others, "The Emerging Need for Family-Centric Initiatives for Obtaining Consent in Personal Genome Research" (2014) 6 Genome Medicine 118.

39 National Centre for Indigenous Genomics (NCIG), "Engagement & Consent" (*NCIG*, March 30, 2015) http://ncig.anu.edu.au/ncig-collection/engagement-consent accessed October 23, 2017.

legal standards for consent, in a world where data protection laws are in flux. Researchers and ethics committees can have confidence that this approach provides a mechanism for participants to exercise their legal rights to choose how their samples and data are used.

5.2 Communication, Choice, Engagement

5.2.1 Long-term engagement

The challenges which biobanks face in terms of maintaining participant engagement over an extended period, particularly to enhance an initial data set by collecting further data or samples, may be resolved through the application of a Dynamic Consent approach. The existence of an online platform which participants can access at their convenience provides for involvement over the lifetime of the research. This has particular benefits for long-term projects examining such things as chronic illness and links between lifestyle and genes, as it enables researchers easily to recontact participants "to discuss expanding their consent or to convey macro-level study results"[40]—a capability which research has indicated may favorably affect willingness to participate.[41] For example, the RUDY study used its inbuilt Dynamic Consent tool to recruit unaffected family members to the project,

> in order for their data, with the agreement of both parties, to be linked, to provide valuable comparison data. Using a traditional consent process, this could have taken around 6 months to complete … Dynamic Consent allowed for the reconsent process to be completed within a two-week period, as participants were sent an email notification, which directed them to the new section of the Dynamic Consent page on their personal profile.[42]

This capacity to expand the scope of decisionmaking about research to include family members, using new technologies, is an important area for future development of Dynamic Consent.[43] The need for long-term data collection has typically been viewed as a risk to successful project completion; Dynamic Consent may help to mitigate this risk by enhancing involvement and retention. There may also be supplemental advantages where research is integrated with clinical

---

40 Whitley, Kanellopoulou, and Kaye, *supra* at note 6, 241.

41 Rhydian Hapgood, Chris McCabe, and Darren Shickle, "Public Preferences for Participation in a Large DNA Cohort Study: A Discrete Choice Experiment" http://eprints.whiterose.ac.uk/10939/1/HEDS_DP_04-05.pdf accessed April 20, 2018.

42 Teare and others, *supra* at note 34, 819.

43 Jusaku Minari and others, "The Emerging Need for Family-Centric Initiatives for Obtaining Consent in Personal Genome Research" (2014) 6 Genome Medicine 118.

care; Wee and others have highlighted the potential for these new mechanisms "to strengthen long-term therapeutic relationships with patients."[44]

5.2.2 Real and effective choice

Dynamic Consent also upholds participant autonomy by enabling true choice as to privacy preferences, and greater transparency in terms of how people's data and samples are being used. Utilizing this consent methodology, people contributing to biobanks will no longer be "restricted to static, one-off, unchanging or time-consuming ways [of giving consent] that permit only single or very limited abilities to indicate or modify their preferences."[45] They can make decisions about participation and privacy and can track and change those decisions; and they can state how and in what circumstances they are to be contacted regarding the specified research or other related projects. An example is the interdisciplinary EnCoRE (Ensuring Consent and Revocation) project, which brought together partners to develop IT solutions giving effect to Dynamic Consent that aimed to "make giving consent as reliable and easy as turning on a tap, and revoking that consent as reliable and easy as turning it off again."[46] Systems-based linkage of consent preferences to data items can ensure that choices made by participants are honored.[47] Researchers can feed back study-level results easily because the communication pathway with participants remains open throughout each project. Furthermore, a Dynamic Consent approach permits greater tailoring, both to individual participant preferences for involvement, and to the needs of specific research projects over time.

5.3 Benefits to Research

The advantages that Dynamic Consent can deliver for participants in biobank research platforms, as outlined within this chapter, are mirrored in gains for the research activity and researchers involved. The characteristic complexity and expense of both initial and follow-up recruitment and consent processes create risks for research projects and for the sustainability of biobank infrastructures. Cross-border collaboration is particularly challenging if consent processes

---

44 Richman Wee, Mark Henaghan, and Ingrid Winship, "Dynamic Consent in the Digital Age of Biology: Online Initiatives and Regulatory Considerations" (2013) 5 Journal of Primary Health Care 341, 344.

45 *Ibid.*

46 EnCoRe, "Ensuring Consent and Revocation" www.hpl.hp.com/breweb/encoreproject/index.html accessed November 22, 2017.

47 Whitley, Kanellopoulou, and Kaye, *supra* at note 6, 234.

at different sites are mismatched.[48] As outlined earlier, the introduction of changes to the scope of a piece of research, or a new study using existing tissue samples and data, could raise ethical concerns and see the need for reconsent processes to be implemented, involving significant time and financial costs. Being able to overcome these challenges through the incorporation of a Dynamic Consent tool, which can be easily adapted to changing circumstances, will likely have flow-on benefits for the economic sustainability of biobanks, since their contents can be accessed and added to more readily, and projects can be established and expanded more efficiently.

6 CONCLUSION

Biobanks around the world have to demonstrate their benefits to research in order to ensure continued funding, support, and long-term sustainability. This is difficult to achieve in an environment where there are changing legal and regulatory requirements as well as social expectations. Digital technologies are used in many areas of life, such as shopping and banking, to provide more efficient ways to engage and communicate with consumers. Dynamic Consent has been developed in medical research to improve communication and engagement with patients and research participants. This has the potential to address many of the practical concerns of consenting and reconsenting over time that have been problematic for biobanks using paper-based systems, as well as being able to deal with the different consent requirements of different jurisdictions. It also addresses the issues of transparency and accountability and enables participants to see the benefits of their altruism and society's investment in research. While the benefits of Dynamic Consent do not address all the issues associated with biobank sustainability, it can assist biobanks to become more resilient and responsive to changing circumstances.

---

48 *Ibid* at 237.

8. Generating trust in biobanks within the context of commercialization: can Dynamic Consent overcome trust challenges?

Esther van Zimmeren[1]

1 SETTING THE SCENE

A growing number of studies are highlighting the importance of trust within the context of biobanking.[2] Trust has been consistently identified as an impor-

[1] The author would like to thank the participants of the "Global Genes Local Concerns" conference at the University of Copenhagen on March 16, 2017 for their interesting feedback.

[2] See for instance: Keymanthri Moodley and Shenuka Singh, '"It's all about trust": reflections of researchers on the complexity and controversy surrounding biobanking in South Africa' (2016) 17 BMC Medical Ethics 57; Dianne Nicol et al, "Understanding public reactions to commercialization of biobanks and use of biobank resources" (2016) 162 Social Science & Medicine 79; Virginia Sanchini et al, "A Trust-based pact in research biobanks: from theory to practice" (2016) 30 Bioethics 260; Casey Lynnette Overby, "Prioritizing approaches to engage community members and build trust in biobanks: a survey of attitudes and opinions of adults within outpatient practices at the University of Maryland" (2015) 5 Journal of Personalized Medicines 264; Christine Critchley et al, "The impact of commercialisation and genetic datasharing arrangements on public trust and the intention to participate in biobank research" (2015) 18 Public Health Genomics 160; Christine Critchley et al, "Predicting intention to biobank: a national survey" (2012) 22 Eur. J. Public Health 139; Peter Dabrock, Jochen Taupitz, and Jens Ried (eds), *Trust in Biobanking: Dealing with Ethical, Legal and Social Issues in an Emerging Field of Biotechnology* (Springer 2012); Alice K. Hawkins and Kieran O'Doherty, 'Biobank governance: a lesson in trust' (2010) 29 New Genetics and Society 311; Hazel Thornton, "The UK Biobank project: trust and altruism are alive and well: a model for achieving public support for research using personal data" (2009) 7 International Journal of Surgery 501; Alastair V. Campbell, "The ethical challenges of genetic databases: safeguarding altruism and trust" (2007) 18 King's Law Journal 227; Mats G. Hansson, "Building on relationships of trust in biobank research" (2005) 31 J. Med. Ethics 415 and Richard Tutton, Jane Kaye, and

tant factor in predicting the attitude and intention of potential participants and the public at large. Biobanking's ethics and governance framework, and in particular issues of consent and the right of withdrawal, will have important implications for securing (public) trust and long-term support, on which the success of many biobanks depends. Moreover, private funding and partnerships with publicly funded biobanks, access to biobank resources, downstream commercialization of products, benefit sharing, and patenting are particularly sensitive issues in terms of building trust.[3] Therefore, in order to minimize the erosion of trust that may accompany commercialization, biobanks should at least provide transparent information regarding the prospects of commercialization and any strategies that might be adopted in this area throughout the lifecycle of a biobank.[4] Hereinafter, I group these concerns under "commercialization issues," wrapping together different questions related to the for-profit considerations in biobanking.

Commercialization issues are especially prominent within the context of "translational research" or "translational medicine." Since the completion of the Human Genome Project, a lot of attention has been given to the translation of genetic research and evidence-based practices into routine health care, including new diagnostic protocols or assays and sometimes even new treatments. However, the actual translation of research results encounters many difficulties along the way: the attrition rates are very high, and it generally takes decades before success is achieved.[5] It has been argued that biobanks can

---

Klaus Hoeyer, "Governing UK biobank: the importance of ensuring public trust" (2004) 22 Trends in Biotechnology 284.

[3] Nicol (2016) 79 and Tim Caulfield et al, "A review of the key issues associated with the commercialization of biobanks" (2014) 1 Journal of Law and Biosciences 94.

[4] Cf Caulfield (2014) 94. This is also reflected in numerous international guidelines and recommendations: see HUGO Ethics Committee, Statement on Human Genomic Databases (2002) www.hugo-international.org/Resources/Documents/CELS_Statement-HumanGenomicDatabase_2002.pdf accessed June 19, 2018; Organization for Economic Co-Operation and Development, *OECD Guidelines on Human Biobanks and Genetic Research Databases* (OECD Publishing 2009) www.oecd.org/sti/biotech/44054609.pdf accessed June 19, 2018; European Society of Human Genetics, "Data storage and DNA banking for biomedical research: technical, social and ethical issues" (2003) 11 Eur. J. Hum. Genet. S8.

[5] See for instance: Bryn Lander and Janet Atkinson-Grosjean, "Translational science and the hidden research system in universities and academic hospitals: a case study" (2011) 72 Soc. Sci. Med. 537–544; Martin Wehling, "Translational medicine: can it really facilitate the transition of research 'from bench to bedside'?" (2006) 62 Eur. J. Clin. Pharmacol. 91–5; John P. Ioannidis, "Materializing research promises: opportunities, priorities and conflicts in translational medicine" (2004) 2 J. Transl. Med. 5.

have a pivotal role in translation and advancing public health.[6] Aside from the results of biobank projects that may ultimately improve medicine, the research process itself can have beneficial effects on the delivery of high-quality health care through the establishment of a research infrastructure and its associated collection practices.[7] The actual processes and tangible care benefits for research participants may already be an important component in safeguarding the trust of biobank donors.[8] Nonetheless, to get the results "from the bench to the bedside," commercial involvement is vital. Empirical research in several jurisdictions has revealed that commercialization issues affect public support and trust in the context of biobanking.[9] Translational medicine, biobanks, commercialization issues, and trust are, thus, closely intertwined.

Interestingly, not all authors who refer to the relevance of trust clarify what type or form of trust they are taking into consideration. As clarified in the extensive trust literature from various disciplines (for example economics, philosophy, management studies, social psychology, organizational sociology), the concept of trust is very complex.[10] It can take many different forms, and in particular situations, several forms of trust can be combined. To complicate the conceptualization of trust, depending on the stream of literature or research discipline, trust researchers tend to use slightly different terminology. It would, hence, be helpful for every study or publication that employs the concept of trust to clarify what the author means by "trust" (for example, interpersonal

---

6 Jennifer R. Harris et al, "Toward a roadmap in global biobanking for health" (2012) 20 Eur. J. Hum. Genet. 1105–11.

7 Conor M.W. Douglas and Philip Scheltens, "Rethinking biobanking and translational medicine in the Netherlands: how the research process stands to matter for patient care" (2015) 23 European Journal of Human Genetics 736.

8 *Ibid.*

9 For instance: Nicol (2016) 79; Tim Caulfield et al, "Biobanking, consent and control: a survey of Albertans on key research ethics issues" (2012) 10 Biopreservation and Biobanking 433; Andrew Webster et al, *Public Attitudes to Third Party Access and Benefit Sharing: Their Application to UK Biobank—Final Report to the UK Biobank Ethics and Governance Council,* Science and Technology Studies, University of York; Gillian Haddow et al, "Tackling community concerns about commercialization and genetic research: a modest interdisciplinary proposal" (2007) 64 Social Sciences & Medicine 272.

10 Some landmark papers on trust can be found here: Reinhard Bachmann and Akbar Zaheer (eds), *Landmark Papers on Trust*, The International Library of Critical Writings on Business and Management series, Vols I and II (Edward Elgar 2008). Some other interesting handbooks on trust include Ellie Shockley et al (eds), *Interdisciplinary Perspectives on Trust: Towards Theoretical and Methodological Integration* (Springer 2016); Reinhard Bachmann and Akbar Zaheer (eds), *Handbook of Advances in Trust Research* (Edward Elgar 2013); and Reinhard Bachmann and Akbar Zaheer (eds), *Handbook of Trust Research* (Edward Elgar 2006).

trust or organizational trust) and who are the trusting parties (for example, the general public, participants, researchers, funders). Therefore, in Section 2 the current contribution provides a short description of the concept of trust that is used within. Clearly, the aim of that description is not to give an exhaustive overview of available theories of trust, but rather to clarify the starting point for this chapter. Moreover, an attempt is made to tailor the trust conceptualization to the context of biobanks for a better understanding of how trust may play a role in different relations and at different levels.

Not only is trust a complex, heterogeneous concept; biobanks are also highly heterogeneous. Biobanks can be defined as repositories of biological materials linked with genetic, genealogical, personal, and health information that can be used for a variety of research purposes to improve the health of individuals and populations.[11] Biobanks may range from small collections of disease-specific blood and tissue samples to large-scale general population-based collections of blood exceeding a million samples. Challenges will be very different depending on the type of biobank, the type of collected samples, the institutional base, and the objectives and mission. This means that the "trust challenges" that will be listed in Section 3 of this contribution will need to be tailored and adapted to the particular needs and features of a specific biobank. Nonetheless, this chapter is based on the assumption that it is possible to provide some more general and yet useful insights related to biobanks, trust, and Dynamic Consent similar to what other authors have done in the legal literature on biobanks and informed consent.

The requirement to obtain informed consent is a foundational principle of medical ethics and international data protection law. It is reflected in all the major ethical and legal declarations and conventions, such as the Nuremberg Code (1946),[12] the World Medical Association's Declaration of Helsinki (1964),[13] and the Convention on Human Rights and Biomedicine of the European Council (1997).[14] It is also restated in many national legal

---

[11] See for instance: Organisation for Economic Co-operation and Development (OECD), *Creation and Governance of Human Genetic Research Databases* (OECD Publishing 2006).

[12] *Nuremberg Code, Trials of War Criminals Before the Nuermberg [Nuremberg] Military Tribunals Under Control Council Law No 10*, vol. II, Washington, DC, U.S. Government Printing Office, 1949–53, 181–3.

[13] WMA, The World Medical Association, *WMA Declaration of Helsinki—Ethical Principles Involving Human Subjects*, www.wma.net/policies-post/wma-declaration-of-helsinki-ethical-principles-for-medical-research-involving-human-subjects accessed May 21, 2018.

[14] Council of Europe, *Convention for the Protection of Human Rights and Dignity of the Human Being with regard to the Application of Biology and Medicine: Convention on Human Rights and Biomedicine*, Oviedo, April 4, 1997, ETS No 164, available at

documents and national and international guidelines. It crystalizes the principles of self-determination and autonomy of patients and research subjects ("participants") and is a way to ensure respect of personal integrity and human dignity. There is, however, an ongoing debate over the most appropriate *mode* for gaining consent from participants that contribute tissue samples and data to biobanks, which would uphold strict ethical standards and enable autonomous decisionmaking.[15] Traditionally, a distinction is made between "broad consent" and "specific consent." Broad consent allows for future research which has not been specified at the time of consent, but which is generally still subject to some content and/or process restrictions. The terms under which broad consent can be considered "informed consent" are controversial and have been extensively addressed in the literature.[16] Consent to specific research purposes ("specific consent") is a rather common approach, but its drawback is that it can limit, or even prevent, certain types of populational or longitudinal research projects. Many authors consider specific consent too difficult to apply in biobank research, where biological sample collections are built as resources for multiple research purposes.[17] Some authors have gone beyond broad consent to suggest so-called blanket consent, that is, consent to an unlimited range of research uses, as a strategy to facilitate research.[18] More recently, metaconsent, which enables participants to express their specific preferences regarding the type of consent they want to give for a specific type of research, has been described as a solution that may contribute to people's willingness to participate in biobank research (for example, broad consent for research on biological samples and specific consent for research by companies).[19] As yet, no international consensus exists regarding the optimal approach for biobanks to balance the interests of donors and researchers.

For a long time, the actual *method* of obtaining consent involved a one-time paper-based written consent. However, such mechanisms are rather expensive

www.coe.int/en/web/conventions/full-list/-/conventions/treaty/164 accessed May 21, 2018.

[15] See also Jane Kaye and Megan Prictor, "Dynamic consent and biobanking—a means of fostering sustainability?" Chapter 7 in this book.

[16] See for instance Björn Hoffmann, "Broadening consent and diluting ethics?" (2009) 35 *Journal of Medical Ethics* 125.

[17] Mast G. Hansson et al, "Should donors be allowed to give broad consent to future biobank research?" (2006) 7 *Lancet Oncology* 266.

[18] Björn Hoffmann, "Broadening consent and diluting ethics?" (2009) 35 *Journal of Medical Ethics* 125.

[19] Thomas Ploug and Søren Holm, "Going beyond the false dichotomy of broad or specific consent: a meta-perspective on participant choice in research using human tissue" (2015) 15 *American Journal of Bioethics* 44.

to administer and not always comprehensible and satisfactory for participants.[20] Increasingly, it is being proposed to use IT-based means to capture consent, through methods such as electronic, computable, or Dynamic Consent.[21] Electronic consent is a method of obtaining and documenting consent using electronic means (such as a computer or smartphone). The benefit of electronic consent relates to its potential to improve efficiency and potentially increase recruitment.[22] Computable consent, or machine-readable consent, goes a step further than electronic consent in automating and streamlining the consent process, as it aims to provide an algorithm to capture the specific consent requirements and limitations.[23] It is aimed at bridging the gap between a static document and the data it contains (for example with regard to restrictions on research use). Dynamic Consent is a tool that takes due account of the interests of participants, enabling them to actively engage in an informed, voluntary, and ongoing manner. It provides a personalized communication interface to enable greater participant engagement and places the participant at the center of the decisionmaking process, allowing continuous interactions over time.[24]

Several authors believe that Dynamic Consent could contribute to "public trust."[25] It would not only serve to streamline, accelerate, and facilitate longitudinal research in the interest of research, but would also respect the international research ethics framework, and would use technological means to allow participants to be actively engaged. The main objective of the current chapter is to explore the extent to which Dynamic Consent could be a more sustainable tool for enabling informed consent, increasing transparency and safeguarding trust by enabling dynamic follow-up, information, discussion, and engagement. In this respect commercialization issues are also explicitly covered, as they are widely known to be one of the main causes of decreased trust among participants. This chapter's contribution to the already extensive literature on biobanks and the growing literature on trust and biobanks relates

---

20 See Kaye and Prictor.

21 Anne Tassé and Emil Kirby, "Is written informed consent outdated?" (2017) 27 European Journal of Public Health 195.

22 Christian M. Simon et al, "Traditional and electronic informed consent for biobanking: a survey of U.S. Biobanks" (2014) 12 Biopreservation and Biobanking 423.

23 Tassé and Kirby (2017) 195.

24 Kaye and Prictor and Jane Kaye et al, "Dynamic Consent: a patient interface for twenty-first century research networks" (2015) 23 European Journal of Human Genetics 141.

25 Renske Broekstra et al, "Written informed consent in health research is outdated" (2016) 27 European Journal of Public Health 194; Hawys Williams, "Dynamic consent: a possible solution to improve patient confidence and trust in how electronic patient records are used in medical research" (2015) 3 JMIR Med Inform doi:10.2196/medinform.3525 and Kaye et al (2015) 141.

to the conceptualization of trust that is being offered and the link that is made to Dynamic Consent. It is hence more theoretical in nature, but substantiated by empirical results drawn from the available empirical literature on biobanks and trust.

As noted, in Section 2 the chapter explores a more advanced conceptual framework of trust for biobanks. Section 3 identifies a list of "trust challenges," a nonexhaustive "wishlist" of items that are considered important for sustaining trust in biobanks. Section 4 evaluates the extent to which Dynamic Consent could deal with some of those trust challenges. Section 5 concludes with some final thoughts and issues that require further research.

2 CONCEPTUAL TRUST FRAMEWORK

In their renowned review paper on the issue of trust, Rousseau and colleagues provide a cross-disciplinary view of trust and a general definition of trust.[26] Their definition is nowadays used widely by researchers from many different disciplines, and clarifies that trust is a "psychological state comprising the intention to accept vulnerability based upon positive expectations of the intentions or behavior of another."[27] This definition clearly reflects that trust especially arises in situations of uncertainty and interdependence. Trust is not relevant if actions can be undertaken with complete certainty and without risks.[28] The suspension of uncertainty, the so-called "leap of faith", and the acceptance of vulnerability by the trustor are the essence of the concept of trust.[29] Uncertainty as to whether the trustee will act appropriately is the source of risk.[30] Risk is thus the first condition for trust to emerge. The second condition for trust to arise is interdependence: the interests of one party cannot be achieved without reliance upon another party. Trust thus relates to a particular connection between the parties, a relation that consists of three parts: A (trustor) trusts B (trustee) to do X.[31]

Without being confident that the trustee will display some competence and is committed to doing X, the trustor will not be inclined to trust. This compe-

---

[26] Denise M. Rousseau et al, "Not so different after all: a cross-discipline view of trust" (1998) 23(3) Academy of Management Review 393.

[27] *Ibid* at 395.

[28] J. David Lewis and Andrew Weigert, "Trust as a social reality" (1985) 63 Social Forces 967.

[29] Guido Möllering, *Trust: Reason, Routine and Reflexivity* (Emerald Publishing 2006).

[30] Rousseau et al (1998) 395.

[31] Russel Hardin, "Trust in government," in Valerie Braithwaite and Margaret Levi (eds) *Trust and Governance* (Russell Sage Foundation 1998) 9.

tence and commitment reflect the "trustworthiness" of B. Trust is an "attitude," whereas trustworthiness is a "property." Trust and trustworthiness are therefore distinct in a way, although ideally those who are trusted are trustworthy, and those who are trustworthy will be trusted. For trust to be warranted (that is, well grounded), parties must be trustworthy. Moreover, trust is not static; it is a dynamic phenomenon that changes over time. Trust can be built and can increase; it can decline and resurface.[32] Trust often emerges through repeated interactions in which parties exchange, take risks, and fulfill positive expectations, strengthening their willingness to continue to rely upon each other. Trust is important in many different ways in our day-to-day life, including through the enablement of cooperative behavior.[33]

Trust can take different *forms* in different relations. For instance, whereas in certain relations trust results from a calculated weighing of benefits and costs (calculus-based trust), in personal relations it might be a more emotional response based on attachment and identification (identity-based trust), capacity (capacity-based trust), or goodwill (goodwill-based trust). These different forms of trust ensue at *different levels*: micro, meso, and macro.[34] Trust can exist at a *micro level*, the interpersonal level, where the "face work" between the trustor and trustee takes place.[35] In the biobank context, this trust level relates, for instance, to the relation between an individual participant (A, trustor) and a researcher (B, trustee), or between a patient (A, trustor) and a doctor/nurse (B, trustee), where B informs A about the biobank (for example, objectives, focus, consent, sampling, and use). Trustors assess the trustworthiness of the trustee based on what they know about the personal predispositions of the trustee, for example, level of education and competence, expertise regarding the biobank/consent procedure/research/disease and so on, communication skills, and commitment. The *macro level* focuses on organizations or systems. An example in the translational medicine context is public trust in the

---

32 Rousseau et al (1998) 395.

33 Cf Diego Gambetta (ed.), *Trust: Making and Breaking Cooperative Relations* (Basil Blackwell 1988).

34 Most authors refer to the micro level for interpersonal trust (for example, trust in a doctor), the meso level for organizational trust (for example, trust in a public hospital) and the macro level for systems trust (for example, trust in the public health system). I propose a slightly different distinction here. My rationale for doing so is that I find it less interesting to distinguish the mere "level" of the trustee and prefer to emphasize the link between interpersonal trust (micro) and organizational trust (macro), on the one hand, and the "external source" of trustworthiness of the trustee (third parties) (meso), on the other. Therefore, I classify both organizational trust and systems trust at the macro level and I "reserve" the meso level to information about the competence and commitment of the trustee based on third party information or experiences.

35 Anthony Giddens, *The Consequences of Modernity* (Polity 1990).

biobank (organizational trust) or in the country's health system (systems trust). At this level, trustors make themselves vulnerable to the actions of the trustee based on what is known about the organizational formal and informal institutions: the applicable legal framework (national, regional, international) and guidelines (local, national, regional, international); the governance framework; information, consultation, and consent processes; and transparency.

These levels of trust are different, because their focal object varies.[36] Within the context of biobanks they are also closely linked, as the researcher or doctor (micro level trustee) can be regarded as a representative of the biobank (macro level trustee). The literature on the link between the interpersonal and the organizational trust is still developing and it is not yet completely clear how exactly these levels of trust are connected and how they interact.[37] According to Kroeger, the key author regarding this interaction, persons who interact with the outside world on behalf of the organization (such as researchers, doctors, nurses)—the so-called boundary spanners—may generate trust. However, as they represent the organization, a transfer from interpersonal to organizational trust can occur if the representative's conduct is viewed as typical of the organization.[38] And conversely, organizational characteristics may facilitate trust in the individual representative when that individual is not yet known by the trustor. In my opinion, aside from the relationship between A and B, the trustor and trustee, a lot of "space" exists that may influence the likelihood and level of trust. In order to acknowledge the relevance of that space, I distinguish a third level, the *meso level*, that does not directly relate to the characteristics or conduct of the trustee and the trustor's perceived trustworthiness of the trustee, but represents "authorities" external to the trust relationship between the trustor and the trustee, such as those providing third party information (such as patient organizations, other patients, (social) media) that influences the cost–benefit analysis and relational considerations (such as historical familiarity, value congruence, experiences). These three levels (see Figure 8.1) are central aspects of the trust conceptualization that I use in the remainder of this paper.

---

36 Akbar Zaheer et al, "Does trust matter? Exploring the effects of interorganizational and interpersonal trust on performance" (1998) 9 *Organizational Science* 141.

37 Frens Kroeger, "Trusting organizations: the institutionalization of trust in interorganizational relationships" (2011) 19 Organization 743.

38 Kroeger (2011) 743, 747.

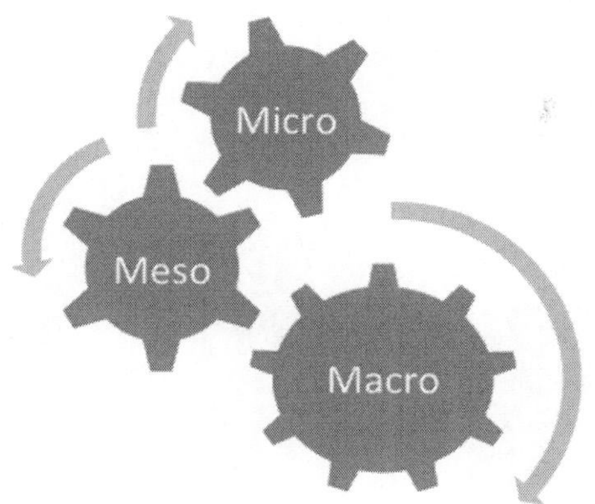

Figure 8.1 *Multilevel trust conceptualization*

3 TRUST CHALLENGES FOR BIOBANKS

Before delving into the various trust challenges for biobanks in more detail, I would like to highlight a particular characteristic of biobanks that adds an additional layer of complexity in generating trust from relevant stakeholders. Biobanks operate as so-called "two-sided platforms".[39] Two-sided platforms serve distinct groups of stakeholders, which need each other in some way. They provide a real or virtual meeting place and facilitate the interactions between groups. In doing this, they essentially act as intermediaries between the two groups and allow them to capture different benefits, to lower transaction costs, and to reduce duplication costs.[40] The two sides of the platform are generally interdependent: the support of a large group of stakeholders on one side of the platform will generally have a positive impact on the interest of the

---

39 Two-sided platforms were first identified in pioneering work by Jean-Charles Rochet and Jean Tirole: "Platform competition in two-sided markets" (2003) 1 J. Eur. Econ. Association 990. Since then a lot of theoretical and empirical research in economics has emerged that has further developed the concept. See for instance David S. Evans and Richard Schmalensee, "Markets with two-sided platforms," Issues in Competition Law and Policy (ABA Section of Antitrust Law 2008), Vol 1, Chapter 28, available at SSRN: https://ssrn.com/abstract=1094820 accessed June 27, 2018. Many important industries are populated by businesses based on two-sided platforms, including more traditional businesses, such as shopping malls, but also most internet-based businesses, such as social networks.

40 Evans and Schmalensee (2008).

group on the other side of the platform. For a biobank to operate as a successful intermediary, it is dependent on the sustainable support of many stakeholders (see Figure 8.2), but in particular that of participants and researchers. This contribution focuses on participants and researchers.

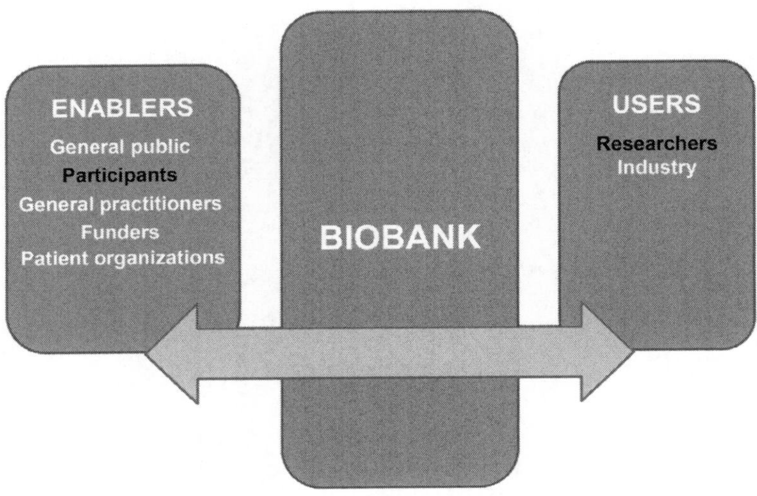

Figure 8.2 *Biobank as a two-sided platform*

The two-sided nature of the platform also has an impact in terms of the trust relationship. Participants give their data and allow samples to be drawn trusting those that are collecting to handle it securely and safely to be used for the years to come. This requires confidence in the processes and systems that have been set up to obtain, handle, and store data and samples, and belief that they will be used for the objectives for which they were obtained. On the other side, the biobank will only be viable if researchers also trust those processes and systems. The information delivered by the biobank to these different audiences needs to be consistent but diversified, as each group's needs, background, level of expertise, and, hence, expectations are different. Yet the level of support, participation, and collaboration is interdependent: if a biobank is not capable of attracting significant numbers of potential participants, the value of collected samples and data will be less valuable for researchers. Similarly, if researchers (publicly) express dissatisfaction with the performance of a biobank, in the long term that may also have an impact on the future engagement of participants and the general level of trust *vis-à-vis* the biobank.

The biobank needs to earn trust; it will not be blindly given. This relates to the micro, meso, and macro levels identified previously.[41] At the micro level, the boundary spanner's[42] personal dispositions (expertise, education, communication) will be vital. For instance, if a potential participant is uncomfortable with the way a staff member is providing information (for instance due to perceived lack of interest/commitment, complex terminology, level of detailed information), this will likely have an impact on the participant's attitude toward the biobank in general. At the meso level, stakeholders will draw on more indirect information and relations. This can include others' (negative/positive) experiences with the biobank, information from third parties such as patient organizations/media, and also more relational considerations, such as past experiences at the hospital with which the biobank is associated and the (private/professional) relationship with doctors/researchers that have worked with the biobank. Finally, at the meso level the cost/benefit considerations of the stakeholders concerned are relevant: for instance, do potential participants care about informed consent and privacy issues or do they mainly consider the cost factor in going through the informed consent procedures? And at the macro level, the stakeholders on both sides of the biobank platform will be influenced by institutional embedding and by the legal and governance frameworks (for example, applicable national law, policies and rules of the biobank, role and tasks of the ethical review board, and the existence, role, and tasks of a stakeholder committee).

The literature has identified several issues that could create challenges in terms of securing trust at these three levels (see also Figure 8.3). These challenges can be grouped as follows: (a) key ethical issues, which takes in (i) informed consent and (ii) right of withdrawal; (b) specific issues, which takes in (i) information and transparency, (ii) governance, and (iii) access and use. These issues cannot necessarily be linked to just one of the three levels. Rather, they cross-cut through the three trust levels, as the levels represent different perspectives that can relate to all the listed issues. For example, clear information will play a role at the micro, interpersonal level in the interaction with the biobank's "boundary spanner," but also at the meso level for information from third parties and at the macro level regarding the legal and governance framework. Moreover, the interests and expectations of the stakeholders on the two different sides of the "biobank platform" might differ for each of these

---

41 See section 2, including Figure 8.1.

42 The person who is considered to be the representative of the biobank: a researcher that is providing information to potential participants; a nurse that is taking a sample; a biobank manager seeking collaborations with researchers or industry or for financial support.

different challenges and the biobank needs to find a way to conciliate those positions in order to keep all stakeholders on board.

3.1 Informed Consent

When a patient or research subject gives informed consent, that can be regarded as an expression of trust,[43] through which the participant declares his confidence in the person that guided him/her through the consent process (micro) or in the biobank (macro). They trust, for instance, that the research will be carried out in accordance with legal and ethical rules,[44] that the risks associated with the research protocol are properly managed, and that their participation is for good reason, with a reasonable potential for gaining new knowledge.[45] However, one might wonder whether that trust is always warranted and to what extent some of the limitations of informed consent will frustrate the proper implementation of informed consent with respect to biobanks.[46] Eyal further argues that informed consent procedures may actually have a negative impact on the interpersonal trust between participants and researchers.[47] For instance, in practice, some participants become suspicious when they are confronted with consent forms. Moreover, in order for informed consent to be meaningful the patient must be able to fully understand the information that is provided, but biobank samples are generally stored for long periods and it is impossible to predict the different types of use that will be relevant over time. Obtaining reconsent for new uses on previously collected samples may not be in the interests of the patient, the donor, the research project, and/or the biobank. Moreover, informed consent in its classical individualistic, static paper-based form is unable to take into account the potential interests of third parties, such as families and genetic relatives. This first challenge ties in with the longstanding debate about broad consent versus specific consent as described in Section 1 of this chapter.

---

43 Mats G. Hansson, "Building on relationships of trust in biobank research" (2005) 31 Journal of Medical Ethics 415.

44 This might concern rules at the international, regional, national, or local biobank level.

45 Hansson (2005) 415.

46 Nir Eyal, "Using informed consent to save trust" (2014) 40 Journal of Medical Ethics 437 and Broekstra et al (2016).

47 Eyal (2014) 437.

3.2 Rights of Withdrawal

The ability to withdraw from participation is an important aspect of consent and it is especially important if a broad notion of consent is employed by a particular biobank. It should, thus, be clear how, under which conditions, and by whom this right can be exercised, and what the consequences of such a withdrawal will be. For instance, can participants state in advance the conditions under which they would like to withdraw, can relatives of a participant have involvement with the right of withdrawal, and is it possible for a biobank to destroy all of the data and the samples collected with regard to a participant who wishes to withdraw?[48] With respect to the latter question, it is generally not possible to destroy all the relevant information due to the backup and audit systems in place and the information that was used by past research projects using the biobank as a resource. Therefore, a clarification should be made that there will be no *further* use of the data by researchers. This would prevent information about the participant concerned from contributing to further analyses, but it would not be possible to remove the data from analyses that had already been done.[49] Participants who are not well informed about these limitations to their right of withdrawal will have certain expectations, if they later invoke this right, that are not warranted. This might have a serious impact on their trust *vis-à-vis* the biobank's "boundary spanner" (micro) who informed the participant insufficiently, and ultimately also on the biobank (macro).

3.3 Information and Transparency

Experience with important biobank projects shows that providing the "right information" in a timely fashion is very important for ensuring the necessary support and participation.[50] When a new biobank is being set up it is hence essential for the governance and ethics council concerned to closely collaborate with a stakeholders' committee in evaluating the information that is offered to the general public, (potential) participants, health professionals, and so on.[51] Feasibility and pilot studies can be helpful in revealing shortcomings

---

[48] Cf Richard Tutton et al, "Governing UK Biobank: the importance of ensuring public trust" (2004) 22 Trends in Biotechnology 284.

[49] UK Biobank Ethics and Governance Council, *Revision of the UK Biobank Ethics and Governance Framework: "No further use" withdrawal option*, June 2007, https://egcukbiobank.org.uk/sites/default/files/Right%20to%20withdraw%20from%20UK%20Biobank.pdf accessed June 19, 2018.

[50] See for instance the UK Biobank project: Thornton (2009).

[51] This applies to information that is made available using different "media": leaflets, website, social media, newspapers, and so on.

that need to be remedied before the project is fully operational.[52] The information should be clear and understandable for a wide public with different levels of education and background. Moreover, the information materials should explain the particular objective and focus of the biobank project, the type of biobank, the process for sampling and collection, and the period during which the biobank will be operational. It should also be clear which type of research will be supported by the biobank. This information should be regularly updated, as it will continuously be influenced by technological and scientific advances. Moreover, participants should understand the extent to which their collaboration will involve particular risks and whether they will get some type of benefit in return (for example, access to individual results), or whether it is more of an altruistic endeavor. Many authors argue that transparency is essential to different forms of trust.[53] If the biobanks' boundary spanner fails to inform the participant properly (micro) and the relevant information is not readily available otherwise from the biobank (macro), participants will likely revert to third parties (meso) in order to assess the trustworthiness of the biobank. Information provided by those third parties may have a strong impact on the biobank's perceived trustworthiness, whereas the actual influence of the biobank on the information that is shared may be very limited. However, the literature on transparency and trust does not give robust insights into the exact relationship between the level of transparency and the trustworthiness of a person or organization, and, in any case, this will be very context-dependent.[54] For instance, the type of information, the quality of the information, the way the information is communicated, the characteristics of the respective trustor and trustee, and so on will all likely have an impact. Therefore, it is necessary to tailor the information to the specific context.[55]

---

[52] Thornton (2009) 502.

[53] See for instance for a review article on transparency, trust and companies: Andrew K. Schnackenberg and Edward C. Tomlinson, "Organizational transparency: a new perspective on managing trust in organization–stakeholder relationships," 42 Journal of Management (2016) 1784.

[54] *Ibid.*

[55] This is also in line with the literature on tailoring information and communication provision on consent processes to their cultural context. Many of these studies focus specifically on particular developing countries and vulnerable patient populations; see for instance Fasil Tekola et al, "Tailoring consent to context: designing an appropriate consent process for a biomedical study in a low income setting" (2009) 3 PLoS Neglected Tropical Diseases e482. However, tailoring of information and communication processes is also highly desirable beyond that context.

3.4 Governance

Consent procedures need to be combined with appropriate internal and external governance mechanisms. The concept of governance has different meanings for different people. Here I use a rather broad but common definition of governance, including both formal institutions and informal arrangements for managing the biobank.[56] The governance framework of the biobank is especially relevant for the macro trust level. In order to sustain the trustworthiness of the biobank, the formal institutions and regimes and informal arrangements should be clarified. It should be clear which bodies are involved in the management of the biobank, what their respective roles and tasks are, and how these bodies interact (that is, "checks and balances"). The various bodies should have clear rules and policies and a clear mandate (including with regard to appointment procedures and decisionmaking processes).[57] In particular, the role and level of control of (private) funders in the governance system should be set out clearly.[58] In addition, prior consultation and communication with communities and interest groups should be safeguarded within the governance framework. Moreover, (external) oversight should be available to hold the biobank accountable for compliance with its rules. Here one could possibly foresee a kind of Biobank Ombuds Office in addition to more "traditional" oversight mechanisms. Quality assurance mechanisms should be firmly embedded within the governance structures. For trust at the micro level, policies on training and reliable, cutting-edge processes for the handling of data and samples, processing, and annotation are necessary. Certification offered by (public or private) third parties could further strengthen trust through the meso level.[59]

---

[56] Commission on Global Governance (ed.), Our Global Neighborhood (Oxford University Press 1995) 2–3: "governance is the sum of the many ways individuals and institutions, *public and private*, manage their common affairs. It is a continuing *process* through which conflicting or diverse interests may be accommodated and cooperative action may be taken. It includes *formal institutions* and regimes empowered to enforce compliance, as well as *informal arrangements* that people and institutions either have agreed to or perceive to be in their interest." (emphasis added)

[57] Cf for instance the suggestions by Richard Tutton, Jane Kaye, and Klaus Hoeyer regarding the Ethics and Governance Framework of the UK Biobank that was under discussion in 2004: Tutton et al (2004) 284, 284–5. They explicitly link these governance issues to UK Biobank's potential to generate "public trust."

[58] Nicol et al (2016) 79, 84.

[59] Cf for instance Matteo Ferrari, "Conveying information, generating trust: the role of certifications in biobanking," in Giovani Pascuzzi et al (eds), *Comparative Issues in the Governance of Research Biobanks: Property, Privacy, Intellectual Property and the Role of Technology* (Springer 2013).

3.5 Sustainability

Biobanks face high risks and are expensive to maintain. Therefore concerns about their long-term sustainability have been growing, triggering various biobank developers to start seeking private investment to facilitate biobanks' long-term sustainability. Moreover, many biobanks do not seem to be able to ensure maximum usage of the available materials. This shows the two-sided platform challenge for biobanks: even when they manage to attract satisfactory numbers of participants willing to contribute to the biobank, for the long-term sustainability of the biobank business model it is also essential that the collection is actually responsive to the needs of researchers and is used for research purposes. A lack of sufficient funding for public or private biobanks may have a dramatic impact on sustainability and may result in bankruptcies. Such bankruptcies raise challenging policy questions, such as what should happen to the participants' samples and data—whether they can be sold, like other types of asset, or transferred to another country, or whether they should instead be destroyed.[60] Given the sensitive nature of medical data, clear policies are required at the macro level. At the moment of the creation of a new biobank the actual risks of bankruptcy and/or merger will likely be downplayed, and policies on this issue will not be considered a priority. Yet it is vital that the full lifecycle of the biobank is taken into consideration in drafting the initial consent documents and the legal and governance framework, as concerns about the sustainability of biobanks are very serious.[61]

3.6 Ownership, Access, and Use

National laws related to ownership and control of human biological materials tend to vary and are sometimes unclear. It would be desirable to clarify such rules at the macro level. Interestingly, empirical research shows that there is no consensus among the general public on the question of who should own human biological material taken, stored, and used for research.[62] However, private entities' involvement in publicly funded biobanks seems to make clarification of this matter imperative.[63]

In order to ensure the general public and participants trust biobanks, it is very important that third parties are prohibited by law from making requests or inquiring at the biobank about genetic or medical information from an

---

60 Caulfield et al (2014) 94.

61 Don Chalmers et al, "Has the biobank bubble burst? Withstanding the challenges for sustainable biobanking in the digital era" (2016) 17 BMC Medical Ethics 39.

62 Caulfield et al (2012) 433.

63 Caulfield et al (2014) 94.

individual, with the exception of specific situations that are described in the legislation, such as certain medical situations or criminal investigations. Therefore, many countries have specific legislation on this topic and biobanks should have clear rules about the actors that have access to genetic or medical information.[64] The macro perspective on trust, consisting of the institutional legal framework surrounding biobanks, is thus vital in this respect.

The two-sided nature of biobanks requires a well-balanced access policy. To protect the rights of research participants, on the one hand, and the professional interests of scientists, on the other, biobanks are increasingly making use of a variety of access agreements. Such access agreements restrict access to key data and samples to researchers who agree to comply with the terms of these access agreements. Often a specific committee will review requests for access, but actual monitoring of compliance and sanctioning of infringers is limited. Nonetheless, even a rather simple access agreement with a limited number of restrictions may deter a substantial number of scientists from applying and undermine the free flow of scientific knowledge.[65] This is another example where the two-sided nature of biobanks leads to additional challenges: in an attempt to protect the rights of participants, biobanks might impose overly restrictive conditions on members of the scientific community. Examples of such conditions could relate to the requirement to sign collaboration agreements with biobanks, to agree to long moratorium periods, to return research results to the biobank, to grant certain rights to potentially emerging intellectual property, or to pay high fees for the provision of samples.[66] Biobank administrators should be careful to use these agreements in a way that ensures support for the rights and interests of those on both sides of the platform. After all, the two sides of the biobank platform are interdependent, and such restrictions may ultimately not benefit research participants either. In this light various authors have proposed alternative arrangements, such as open source and open access models.[67] Guidelines for access agreements can be regarded

---

[64] See for instance for Sweden: Hansson (2005) 417.

[65] Caulfield et al (2014) 94 citing Dov Greenbaum et al, "Genomics and privacy: implications of the new reality of closed data for the field" (2011) 7 PLoS Comput. Biol.

[66] Caulfield et al (2014) 94.

[67] See for instance various contributions in the edited volume by Giovanni Pascuzzi et al (eds), *Comparative Issues in the Governance of Research Biobanks: Property, Privacy, Intellectual Property and the Role of Technology* (Springer 2013): Donna M. Gitter, "The challenges of achieving open source sharing of biobank data" 165; E. Richard Gold and Dianne Nicol, "Beyond open source: patents, biobanks and sharing" 191; Roberto Caso and Rosanna Ducato, "Opening research biobanks: an overview" 209; Mark Perry, "Assessing accessions: biobanks and benefit-sharing" 267.

as a meso level instrument to generate trust among stakeholders at both sides of the biobank.

Transnational collaborations that involve biobanks funded by major (public) research funding schemes, such as the EU's Horizon2020, or by private funders such as the Wellcome Trust are quite common nowadays. Apart from the standardization and interoperability questions involved in such projects, such projects will likely also involve cross-border use of samples and data. In addition, aside from such formal transnational collaborations, cross-border access and use of samples/data is also increasingly requested on a more informal basis. National legislation may restrict such cross-border use for different reasons, or participants may have concerns about the export of samples and/or data. Nicol and colleagues show, for instance, "that trust [was] higher when the researcher accessing the biobank was Australian compared to international."[68] Due to the limited public funding available and the high financial burden involved in translational research, engagement with the for-profit sector is inevitable. Yet there is a large body of theoretical, policy, and empirical research that identifies commercialization issues as one of biobank participants' main concerns with the possibility of affecting their willingness to participate.[69] Empirical research in Australia, Canada, the US, and the UK has shown an inherent distrust of private sector research versus that in the public sector. In particular, the involvement of private research organizations or private funding sources, or a connection between public researchers and pharmaceutical companies, significantly reduces the trust in biomedical research in general, and in biobanks more specifically.[70] It is acknowledged that to get the results "from the bench to the bedside," commercial involvement is vital.[71] Yet, though people might expect "that commercial parties need to be involved with biobanks one way or another," there is "a 'natural prejudice' against involvement of commercial interests in biobanking."[72] Therefore, "appropriate

---

[68] Nicol et al (2016) 79, 82. I would like to highlight in this respect that "this was the case even when the Australian researcher was in the private sector and the international researcher was in the public sector."

[69] Nicol et al (2016) 79, 80; Tim Caulfield et al (2012) 433; Webster et al; Haddow et al (2007) 272 and Campbell (2007).

[70] See Nicol et al (2016) 79, 80, who provide an excellent overview of the available literature on the topic.

[71] Heidi Hörig, Elizabeth Marincola, and Francesco Marincola, "Obstacles and opportunities in translational research" (2005) 11 Nature Medicine 705–8.

[72] Nicol et al (2016) 79, 83. Nicol et al collected this information through empirical research by way of a "deliberative democracy event" focusing on the "hopes and concerns expressed by participants in relation to the potential development of a Tasmanian biobank" and "issues of importance to them." The group deliberations resulted in 17 resolutions across eight issues that they identified as being of importance: consent,

checks and balances need to be in place in order to ensure that commercialization is managed properly so that the public good focus of biobanks is not lost."[73] However, it is not always clear what it is about commercialization that concerns people and which elements are most important in reducing trust.[74] Participants may see the involvement of private funders, the payment of a fee for access to biobank resources, and downstream profitmaking by the biobank as at odds with their initial (altruistic) reasons for and expectations regarding participation.[75]

It is an ethical imperative that biobanks and their boundary spanners are transparent regarding intended commercial use, and that they address participants' concerns related to commercialization.[76] Given the increasingly common use of broad consent by biobanks, the consent given may be insufficient if commercialization issues are not clarified at all at the moment of the biobank's establishment, or only refer vaguely to possible future commercial applications or industry involvement. If commercialization issues are not sufficiently addressed, engaging a private entity after the original consent was given would likely require a demanding reconsent process.[77]

---

privacy, commercialization, benefit sharing, return of results, governance, funding, and community involvement. One of the participants is cited as stating: "Well, we [her small group] were against pharmaceutical companies." Specific concerns were also expressed about patenting and monopolizing research results stemming from the use of biobank resources. See also Caulfield et al (2014) 94.

[73] Nicol et al (2016) 79, 83. One participant in the deliberative democracy event remained resolute, however, stating that "no form of commercialization was acceptable," and two others continued to express concerns despite the attention to checks and balances within the biobank governance system.

[74] Nicol et al (2016) 79, 80 and Caulfield et al (2014) 94.

[75] Caulfield et al (2014) 94 and Nicol et al (2016) 79, 83–4.

[76] Sigrid Sterckx et al, "'Trust is not something you can reclaim easily': patenting in the field of direct-to-consumer genetic testing" (2013) 15 Genetic Medicine 382.

[77] Timothy Caulfield, "Biobanks and blanket consent: the proper place of the public good and public perception rationales" (2007) 18 King's Law Journal 209. For example, broad consent to use samples for a wide range of commercial and noncommercial purposes in the future will probably require the provision of more detailed information about the nature of the commercial activity in cases where a commercial entity is involved.

Key Issues
1. Informed consent procedure
2. Right of withdrawal

Specific Issues
Information & Transparency
1. Clarity information
2. Objective and focus project
3. Research use
4. Duration
5. Risks and benefits (e.g. no individual benefit, access to individual results, financial benefits)
6. Process (e.g. recruitment, sampling, measurements, confidentiality, voluntary, ethical review)

Governance
1. Clear rules and governance system
2. Type of funders and their role in the governance system
3. Bodies (e.g. Biobank Ombuds Office, ethical committee (incl. appointment procedure, decision-making processes))
4. Oversight
5. Quality staff (e.g. training, handling data and samples, processing, annotation)
6. Certification
7. Sustainable business model

Ownership, Access & Use
1. Ownership
2. Parties having access
3. Cross-border use of samples/data
4. Objectives use (especially commercial use)
5. Efficiency process access and use
6. Standardization and interoperability

Figure 8.3 Trust challenges for biobanks

4 DYNAMIC CONSENT AS A TOOL FOR GENERATING TRUST?

At the heart of Dynamic Consent is a personalized, digital communication interface that connects participants and researchers, placing participants at the center of decisionmaking.[78] The core benefit of Dynamic Consent for dealing with the trust challenges listed here is that it allows participants to tailor, update, and adapt their own consent preferences. For biobanks consent has thus become more dynamic, as participants can differentiate their consent in terms of the kind of study (for example, broad consent, with specific consent for private research) and can consent to new projects or modify their consent choices in real time should their circumstances change. The right of withdrawal can be used in a more nuanced way by carefully identifying specific research that the participant finds objectionable instead of a complete withdrawal from the biobank. The technical architecture allows secure encryption of sensitive data and participant consent preferences to travel with their data and samples when they are shared with third parties. In this way participants can be confident that the nuanced specifications and later modifications will take effect, which was much more uncertain with the traditional paper-based system. Traditional paper-based written informed consent is too static and not appropriate for biobanks. Moreover, written informed consent forms may have a limited impact on trust, as trust is more dynamic and typically built though repeated interactions between parties. Dynamic, continuous, tailored, interactive methods of informing and communicating may do more justice to the underlying principles of informed consent by honoring rights of participants and by facilitating them in their decision to participate based on a qualified consent, while being more transparent.[79] Nonetheless, the use of digital tools also has some disadvantages in terms of trust. The use of digital tools will only improve the trustworthiness of the biobank for those participants who do not distrust digitalization and are able to use such tools. This is not an ideal solution for example for potential participants with limited digital access, or those who are disabled or elderly.[80] Therefore, it is essential that sufficient "interpersonal support" is offered for those who do not feel comfortable using digital tools

---

[78] The description of dynamic consent and its potential advantages for participants and researchers is based on Kaye and Prictor; Isabele Budin-Ljøsne et al, "Dynamic Consent: a potential solution to some of the challenges of modern biomedical research" (2017) 18 BMC Medical Ethics https://doi.org/10.1186/s12910-016-0162-9, and Kaye et al (2015).

[79] Kaye and Prictor; Budin-Ljøsne et al (2017); Kaye et al (2015) 141; Broekstra et al (2017); Tassé Kirby (2017).

[80] See also Budin-Ljøsne et al (2017).

(by themselves) and to offer alternatives (including face-to-face meetings) for people who prefer to use traditional consent mechanisms. Moreover, with the introduction of the digital tool the trust level seems to have shifted from the interpersonal, micro level, where the role of the boundary spanner was prominent, to the meso level, where the dynamic trust "relationship" is no longer interpersonal but based on the actual features, functioning, reliability, security, and so on of the digital tool itself. This entails, hence, a move to so-called "digital trust."[81]

Dynamic Consent may also be very beneficial for dealing with some of the other trust challenges for biobanks.[82] This interface facilitates two-way communication to stimulate a more informed, engaged, and scientifically literate participant population. Information that is provided to participants can be presented in a more user-friendly manner, in different languages and using various types of media (for example, text, videos, animation) enabling the participant to select the most appropriate source. The interface would also allow easy modification of contact details, indication of information the participants are interested in receiving, enrollment in new studies, and survey completion. Some digital tools, such as the platform used in the RUDY study,[83] also enable participants to include health and lifestyle data that are less relevant for the research, but which may make it more attractive for participants to use the tool actively and to remain involved in the biobanking. As such options were included at the request of participants, the RUDY study also shows that it is very helpful to involve participants in the design and operation of the platform.[84]

Researchers can provide participants with regular updates on early findings, follow-up studies, and key results. This will likely lead to a more realistic understanding of the research process, to an enhancement of participants' confidence in its transparency and accountability, and to the development of more appropriate expectations among participants. Some participants may further be interested in initiating communication with researchers or other participants, for instance via online discussions, forums, and webinars. Such an interac-

---

[81] A discussion of the emerging literature on digital trust would be interesting in this respect, but is beyond the scope of this chapter. See for instance Zheng Yan and Silke Holtmans, "Trust modeling and management: from social trust to digital trust," in Jonathan Bishop (ed.), *Examining the Concepts, Issues, and Implications of Internet Trolling* (Information Science 2013).

[82] I will not systematically deal with all the listed specific issues but I have selected a number of issues that seem most prevalent within the context of dynamic consent. Most other issues can considered to be covered by the discussion on the information and communication function of the digital dynamic consent platform.

[83] Teare et al (2017).

[84] *Ibid.*

tive use of the platform could positively impact on willingness to continue participation in the biobank. It also emphasizes the interpersonal relationship between participants and researchers, rather than the interaction between the participants and the biobank on the one hand and the biobank and the researchers on the other (two-sided platform). This might indirectly also lead to a more sustainable relationship between the participant and the biobank, but it also requires a continued commitment from researchers. Moreover, there is a risk that participants may start to perceive the researchers who use samples and data from the biobank for their research as representatives, or "boundary spanners" of the biobank. As this might lead to misconceptions and distrust, efforts should be made to ensure that participants fully understand the role and responsibilities of the different actors involved.

The Dynamic Consent interface should be firmly embedded within the governance framework of the biobank and be instrumental for the framework. This means that the interface can be used to provide information on the framework and to explain it in such a way that is understandable for participants with different backgrounds. Moreover, oversight mechanisms should also be competent to review the operation and accountability of the digital tool. The role of participant partnerships within the context of the design and further development of the digital interface has already been highlighted. However, researchers' involvement in the platform design and operation of the interface to tailor the platform to their needs would be desirable too. They might, for instance, appreciate it if the platform could be configured to work with a wide range of laboratory management systems. This could be a task for some kind of stakeholder forum. Certification of the digital tool could further contribute to its trustworthiness.

Researchers will not be able to foresee all opportunities for future research enabled by the biobank or future prospects for commercialization of research findings. Moreover, circumstances may change in terms of ownership and access rules (for example, a biobank sold by a hospital to a company; a change of legal and ethical rules) and new questions related to commercialization, patenting, and downstream product development may come up that were not anticipated at the moment the initial informed consent was required. Specific consent requires participants to be informed appropriately about all such developments in a tailored and interactive way, as described previously. They should be given an opportunity to reconsent, withdraw if they wish to do so, or further differentiate their preferences regarding access and use. However, the management of biobanks would become overly burdensome and unsustainable if this were done through written informed consent. Dynamic Consent enables this process in a flexible and cost-effective way. As the research progresses, participants can continue to be involved in the process, with information provided about the different steps in the research project, provisional results, patent applications, downstream products, benefit sharing to the biobank, and so on.

5 FINAL REMARKS

Dynamic Consent has the capacity to operate as a more sustainable tool for biobank management as it streamlines the recruitment process and enables dynamic follow-up and recontact, information, discussion, and engagement in the interest of both participants and researchers (cf two-sided platform). Participants will no longer be passive human "subjects," but are recognized as active, interested valued participants in this process. The interface could also facilitate many different steps in research projects in which biobanks are involved. It appears to be a tool that could effectively contribute to generating trust among participants and researchers in the biobank.

Examining the role of digital and virtual tools that could contribute to the trust in biobanks is also vital in respect to the increase of transnational collaborations. The empirical research by Nicol and colleagues shows the reluctance of potential participants if nonnational researchers get access to the biobank.[85] It would be worthwhile to further pursue questions of the underlying reasons or assumptions that result in that reluctance. The introduction of digital tools that would enable a better understanding of the relevance of such transnational collaborations, provide information about the actual collaborations, and allow for (virtual) interaction between local participants and foreign researchers seems advisable. For such transnational collaborations, standardization and interoperability are another key challenge, as national legislation and local practices tend to differ. Also in this respect, Dynamic Consent could play a major role in terms of allowing for differentiation in terms of consent processes and other trust challenges, tailoring them to the needs of local participants and creating cost-effective transnational information and communication processes. In this way, there is no need to adopt a new top-down uniform approach to be adopted by all biobanks involved in the transnational collaboration: the digital architecture would ensure that samples and data would only be used in line with the differentiated tailored preferences of the participants. If recontact of participants is required, the digital tool would enable researchers to do so in a much more efficient and sustainable manner.

This chapter relies heavily on the empirical research on biobanks and trust carried out by various excellent international research groups. More empirical research on perceptions, attitudes, and reactions toward Dynamic Consent tools is required to verify the ideas presented here regarding the different trust

---

[85] Nicol et al (2016).

levels. The distinction between the micro, meso, and macro levels that is used in this chapter might help to more clearly identify how to promote trustworthy institutions, interfaces, processes, and practices. The chapter has acknowledged the link between interpersonal and organizational trust, which seems to be quite critical within the context of biobanks. Moreover, it has highlighted the intricate dynamics and potential role of particular persons (boundary spanners or perceived boundary spanners) and third parties in influencing the level of organizational trust in biobanks when they use Dynamic Consent interfaces.

9. Exploitation and vulnerabilities in consent to biobank research in developing countries

Nana Cecilie Halmsted Kongsholm

Biobanks hold great potential for research into a large number of health issues and developments—a potential that is magnified even further when biobanks collaborate internationally. To facilitate international collaboration and overcome the often very large differences between countries' legal and regulative landscapes, many efforts at harmonization of biobank guidelines—particularly with respect to the issue of consent to participation—have been launched. However, such harmonization efforts risk overlooking or glossing over local factors that may compromise consent in specific contexts, particularly when research is conducted in developing countries, which risks enabling exploitative research interactions.

In this contribution I argue that harmonization efforts with a sole focus on consent may overlook two issues: (1) even when there is valid consent, there may still be exploitation (and a focus on consent as the ultimate mark of validity may take attention away from this fact); (2) in concrete research contexts in developing countries, there may be local social and cultural factors that compromise individuals' consent and make them particularly vulnerable to potentially exploitative research interactions.

To illustrate (2) I draw on findings from an interview study conducted with biobank donors in rural Pakistan, carried out under the auspices of the Global Genes, Local Concerns project. Based on these findings I argue that vulnerabilities to exploitation in research are highly dependent on social conventions and customs, and are thus only apparent when due competent attention is paid to such factors. On this basis I further argue that harmonization efforts in international biobank guidelines should be supplemented with appropriate efforts to highlight and accommodate such vulnerabilities in each relevant research context.

1 THE POTENTIAL OF INTERNATIONAL RESEARCH BIOBANKING

While biobanking is not a new phenomenon (the systematic collection of human cells and tissues dates back to the nineteenth century), the past two decades have seen considerable development in its use in the context of genomic research. Here, samples are increasingly used in genome-wide association studies, using large collections of samples to identify biomarkers for disease; in the development of personalized medicine; or to uncover the genetic basis of certain genetic disorders. This recent surge in biobanking activity and research has offered an increased ability and interest in linking biological and genetic data with general information about patients and/or donors, as this allows for the inclusion of demographic information about donors in the study of for example cancers and lifestyle diseases, and in the study of genetic and environmental factors in the etiology of disease.[1] To a great extent, this has been facilitated by the advent of large patient registries and advances in genetics, both of which have dramatically increased the ability to tap the collection of material and its potential uses.

This potential is magnified even further with biobanks operating internationally. There is a still increasing international collaboration between biobanks situated in different countries, and recent years have seen the rise of international biobanking infrastructures, for example BBMRI-ERIC.[2] Such collaborations often involve physical samples and their associated data being exchanged across national borders, which has a number of benefits: it provides researchers with a vastly expanded material for research, to which they otherwise would not have had access; it may provide grounds for comparisons of different nationalities with respect to a specific disorder; and it may provide researchers working on rare diseases not common in their home country with material for research.

---

1 Herbert Gottweis and Kurt Zatloukal, "Biobank Governance: Trends and Perspectives" (2007) 74(4) Pathobiology 206.

2 "Biobanking and Biomolecular Resources Research Infrastructure—European Research Infrastructure Consortium" www.bbmri-eric.eu/BBMRI-ERIC/about-us/ accessed 30 August 2017.

2 HARMONIZATION OF BIOBANK GUIDELINES

This research potential has, however, in some instances been hampered or even halted due to differing and conflicting biobank regulations and guidelines between countries, especially with respect to the issue of informed consent.[3]

2.1 Informed Consent: A Contested Issue in Biobanking

Informed consent is—even on a national level—in itself a contested issue in the context of biobanking. In medical ethics, informed consent is generally held to be an indispensable element of medical research involving human subjects or material. One reason for this is that informed consent is thought to protect individuals' fundamental right to have a say in what happens to their physical being, and to decide, on adequately informed grounds, whether and how their body or body parts and associated information will be used in research. For informed consent to participation in medical research to be ethically and legally valid, participants should be provided with information about the purpose, methods, demands, risks, inconveniences, discomforts, and possible outcomes of the research, and on this basis voluntarily agree to participate in the research.[4]

In the context of clinical research, meeting these requirements is usually rather straightforward: in most cases, researchers have fairly well-defined ideas about the purpose and nature of research; its likely demands, risks, inconveniences, and discomforts; and the possible benefits and/or results hoped to be gained from the research. Hence, it is relatively easy to disseminate this information to research subjects, and on this basis for research subjects to make a decision about participation in the research.

When applied to the biobanking context, however, this ideal of informed consent becomes a notoriously tricky endeavor. Given that the nature and aim of (at least many) biobanks is to store samples for an indefinite duration of time and to allow access for multiple researchers for multiple purposes, it becomes

---

3 Anne Cambon-Thomsen, Emmanuelle Rial-Sebbag, and Bartha M. Knoppers, "Trends in Ethical and Legal Frameworks for the Use of Human Biobanks" (2007) 30(2) European Respiratory Journal 373; Bernice S. Elger and Arthur L. Caplan, "Consent and Anonymization in Research Involving Biobanks" (2006) 7(7) EMBO reports 661.

4 Nir Eyal (ed.), "Informed Consent" (*Stanford Encyclopedia of Philosophy,* 20 September 2011) https://plato.stanford.edu/archives/fall2012/entries/informed-consent/; Wendy Lipworth, Rachel Ankeny, and Ian Kerridge, "Consent in Crisis: The Need to Reconceptualize Consent to Tissue Banking Research" (2006) 36 Internal Medicine Journal 124.

impossible to inform individuals—at least with the specificity required by current standards for informed consent—comprehensively about the research in which they (or, more accurately, their sample) will be participating. Since the content of future research projects cannot always be predicted at the time of sample collection, informing participants about the nature, purpose, risks, benefits, and potential results of such future research uses becomes a complicated, if not impossible, task.

On a national level, biobanks have responded to the challenges regarding informed consent in biobanking in different ways, often based on local legislation and regulations. Notable differences exist between, for example, the approach to consent taken by Europe and the USA,[5] and a 2011 study of 123 European biobanks shows remarkably different consent guidelines even within the continent.[6] This patchwork of frameworks and guidelines for consent, and a lack of international consensus on which guidelines and principles to follow, interferes with international collaboration, which may ultimately delay or halt scientific progress.

As a reaction to the difficulties posed by the great variance of consent requirements and guidelines, many commentators have stressed the urgent need for international harmonization of biobank guidelines,[7] and several such attempts have been initiated.[8] As a potential solution it has been suggested to implement an international scheme of general or broad consent,[9] where any donor consents to a certain framework for future research of certain types.[10]

3 EXPLOITATION: AN OVERLOOKED ISSUE IN BIOBANK RESEARCH IN DEVELOPING COUNTRIES?

These efforts with respect to biobank guidelines will, if successful, undoubtedly facilitate a great amount of research for the benefit and health of the entire world population.

---

[5] Elger and Caplan, *supra* note 3.

[6] Eleni Zika and others, "A European Survey on Biobanks: Trends and Issues" (2011) 14 Public Health Genomics 96.

[7] *Ibid.*

[8] K. Hoeyer, "The Ethics of Research Biobanking: A Critical Review of the Literature" (2008) Biotechnology and Genetic Engineering Ethics 25, 429–52.

[9] Elger and Caplan, *supra* note 3; Bartha M. Knoppers, "Biobanking: International Norms" (2005) 33 Journal of Law, Medicine and Ethics 7.

[10] Kristin S. Steinsbekk, Bjørn Myskja, and Berge Solberg, "Broad Consent versus Dynamic Consent in Biobank Research: Is Passive Participation an Ethical Problem?" (2013) 21(9) European Journal of Human Genetics 897.

However, one issue that they do not address—which is nevertheless paramount in international research guidelines—is the exploitation of research subjects, particularly in developing countries. Exploitation of subjects in clinical research and drug trials in developing countries has been the topic of much recent literature and debate;[11] nevertheless, existing biobank guidelines do not consider this issue to any notable extent. One reason for this may be that the risks and benefits associated with biobanking are negligible in comparison with other types of medical research which have previously been the focus of exploitation debates (for example the Havrix trial in Thailand and the Surfaxin trial in Bolivia[12]). In light of this fact, it may be argued that biobanking as such does not generate any potential for exploitative research interactions.

Nevertheless, a closer look at particular research setups in developing countries involving the collection of material for genomic research reveals that in this context there may, in fact, be grounds for potential exploitation.

3.1 Exploitation in Medical Research in Developing Countries

Much has been written on the topic of exploitative research in developing countries, most commonly oriented toward scenarios where researchers based in wealthy, developed countries conduct medical research on individuals or communities in poorer, developing countries.[13] Such writings usually follow one of two primary strands of thought with regard to exactly what makes the research exchange exploitative.

On the one hand, we have what we may call the *unfair benefits* account.[14] According to this account, research is exploitative when the resulting benefits are unfairly distributed between the participating parties (for example, research subjects on the one hand, and researchers and/or sponsors on the other).

This is the case when research subjects do not receive a fair share of the benefits generated from research (for example, if samples from a certain community are used in the groundbreaking development of a vaccine resulting in large monetary rewards for the researchers, without any proper acknowledg-

---

[11] Jennifer S. Hawkins and Ezekiel J. Emanuel, *Exploitation and Developing Countries—The Ethics of Clinical Research* (Princeton University Press 2008); Ruth Macklin, "Bioethics, Vulnerability and Protection" (2003) 17(5–6) Bioethics 472.

[12] Hawkins and Emanuel *ibid* at 55–61.

[13] Macklin, *supra* note 11; Angela Ballantyne, "'Fair Benefits' Accounts of Exploitation Require a Normative Principle of Fairness: Response to Gbadegesin and Wendler, and Emanuel et al" (2008) 22(4) Bioethics 239.

[14] Ezekiel J. Emanuel and others, "Moral Standards for Research in Developing Countries: From 'Reasonable Availability' to 'Fair Benefits'" (2004) 34(3) Hastings Center Report 17; Segun Gbadegesin and David Wendler, "Protecting Communities in Health Research from Exploitation" (2006) 20(5) Bioethics 248.

ment and/or reimbursement for the involved research participants), or when resulting developments are not made reasonably available for the participants (for example, if samples from the same community are used in the ground-breaking development of the same vaccine, but the vaccine is priced at a level that means the participants will never be able to afford it).

On the other hand, we have what we may call the *invalid consent* account. On this account, individuals or communities in developing countries are at risk of exploitation in research because their consent is compromised or made invalid. This is usually held to be so due to one of two reasons (or, in some cases, both of these).

First, the generally lower level of education that is a characterizing feature of many developing countries, resulting in a lack of understanding of the nature of medical research, may mean that individuals do not have a proper understanding of what exactly they are consenting to—often resulting in therapeutic misconception.[15] Second, many developing countries are marked by widespread poverty and poor medical infrastructure; these factors mean that individuals in such countries suffering from disease and disability may have very limited access to medical care and attention. It is argued this may compromise the validity and voluntariness of individuals' consent to research participation: due to their restricted options, individuals in this position are likely to feel pressure to agree to participate in any type of medical research that they are offered and that they believe will alleviate their situation to even the slightest degree.[16]

It should be noted that this second reading of the invalid consent account has, rightly, been met with some criticism. Critics have raised the point that there is nothing about being faced with a severely limited option set, or extremely dire circumstances, that renders a choice made in this situation invalid or involuntary as such: if my appendix is about to burst, it does not seem right to claim that my choice to go in to surgery is involuntary, or that the situation makes my consent to surgery invalid. Hence, it seems that individuals can indeed make rational and voluntary decisions when faced with a limited option set and/or

---

[15] Therapeutic misconception "occurs when a research subject fails to appreciate the distinction between the imperatives of clinical research and of ordinary treatment, and therefore inaccurately attributes therapeutic intent to research procedures": Charles W. Lidz and Paul S. Appelbaum, "The Therapeutic Misconception: Problems and Solutions" (2002) 40(9) Medical Care; Supplement: V-55–V-63.

[16] Christine Grady, "Vulnerability in Research: Individuals with Limited Financial and/or Social Resources" (2009) 37(1) Journal of Law, Medicine & Ethics 19; Nicholas A. Christakis, "The Ethical Design of an AIDS Vaccine Trial in Africa" (1988) 18(3) Hastings Center Report 31; George J. Annas and Michael A. Grodin, "Human Rights and Maternal-Fetal HIV Transmission Prevention Trials in Africa" (1988) 88(4) American Journal of Public Health 560.

dire circumstances. Notwithstanding such criticism, this account still pervades much literature on exploitation in research in developing countries; hence it shall not be disregarded in the present discussion.

4 EXPLOITATION WITH CONSENT AND BENEFITS: A THEORETICAL VIEW

So far we have seen that, according to the literature, exploitation in research in developing countries occurs when the research participants do not benefit fairly, or when their consent is invalid, or, in unfortunate cases, even both.

Nevertheless, it seems that even in a situation where both parties benefit and there is valid consent, there may still be an issue of exploitation. In a book written in 1996, Alan Wertheimer accounted for the concept of *mutually advantageous and consensual exploitation*. To illustrate this notion, Wertheimer offers the following case (entitled *Snow Shovel*):

> An unexpected blizzard hits an area and people rush to the hardware store to buy a shovel. The hardware store owner, A, sees the opportunity to make an abnormal profit and raises the price of a shovel from $15 to $30. Both parties gain. But B feels exploited because B gains less (or pays more) than B thinks is reasonable.[17]

Here, it is the case that B both *benefits* and *gives valid consent*. But it does not seem far-fetched to say that B is still *exploited*. This should prompt us to take a step back and consider a general and theoretical view of what exploitation is. For present purposes, I offer the following very general conception of exploitation: *A exploits B when A takes (unfair) advantage of B's situation, to make a(n) (unfair) benefit for himself.*

Exploitation understood according to this conception may very likely occur in situations of power asymmetry, where one party is *vulnerable* to the other and what they may offer, and where this offer is characterized by an *unfair benefit profile* to the advantage of the exploiter. Hence, we should understand vulnerability to exploitation as being in a position in which one is more likely to consent to a transaction with an unfair benefit profile than one otherwise would be.

Armed with this general understanding of exploitation and vulnerability, let us take a closer look at what may make someone vulnerable to exploitation in a research setting in developing countries.

[17] See Alan Wertheimer, *Exploitation* (Princeton University Press 1996) 22.

4.1 How Should We Understand Vulnerability to Exploitation in Research?

Several sets of guidelines for ethical research on human subjects address the issue of vulnerable groups, and stress that such groups—due to certain characteristics and circumstances—are more easily exploited in research and thus require extra, special protections. As one example, guideline 13 of the Council for International Organizations of Medical Sciences' (CIOMS) 2002 guidelines states, "Vulnerable persons are those who are relatively (or absolutely) incapable of protecting their own interests. More formally, they may have insufficient power, intelligence, education, resources, strength, or other needed attributes to protect their own interests."[18]

As such, this notion of vulnerability can be viewed as an extension of the invalid consent account of exploitation detailed previously, particularly with respect to the attributes of insufficient education and resources: lack of knowledge with respect to medical research and/or extreme poverty may more easily push individuals with these attributes into research participation compared to individuals without them. Even though it may be true of a few select cases that this has, in fact, led to exploitation in some form, as we have seen, it is a fallacy to equate necessity with openness to exploitation in the form of invalid consent.

The CIOMS guidelines and similar documents (for example the Belmont report of 1979) have rightly been met with criticism regarding their approach to vulnerability.[19] According to critics, such guidelines take what has been termed a "labeling" or "subpopulation" approach to vulnerability, defining and labeling certain groups as vulnerable in virtue of specific characteristics or attributes, and claiming that these characteristics and attributes *in themselves* make individuals vulnerable to exploitation, where this vulnerability is conceived of as a *fixed* state (according to guidelines, this includes, for example, the illiterate, uneducated, and resource-poor). However, this approach overlooks the reality that individuals, even though they may hold one or more of

---

[18] Council for International Organizations of Medical Sciences, *International Ethical Guidelines for Biomedical Research Involving Human Subjects* (2002).

[19] See for example K. Kipnis, "Vulnerability in Research Subjects: A Bioethical Taxonomy," in National Bioethics Advisory Commission [NBAC], *Ethical and Policy Issues in Research Involving Human Participants. Volume II: Commissioned Papers* (G1-G13) (Rockville, MD 2001); Samia A. Hurst, "Vulnerability in Research and Health Care" Describing the Elephant in the Room?" (2009) 22(4) Bioethics 191; Florencia Luna, "Elucidating the Concept of Vulnerabilities: Layers Not Labels" (2009) 2(1) International Journal of Feminist Approaches to Bioethics 121; Grady, *supra* note 16.

the attributes associated with vulnerability as detailed above—are not vulnerable to everyone and at all times, but usually only to some people and in some situations. Rather, vulnerability is highly contextual and situational—a point which is not least relevant in research contexts, as noted by Levine and colleagues, who highlight: "[A]n individual's needs for special protection depend not solely on that person's inclusion in a group, but importantly on the particular features of the research project and the environment in which it is taking place."[20]

Taking a labeling or subpopulation approach to vulnerability in research situations is problematic on both theoretical and practical levels: merely listing groups as vulnerable in virtue of certain characteristics or attributes does not shed any light on *how and why* these make individuals vulnerable to exploitation, and hence does not offer any practical guidance as to what can and should be done to accommodate it.

4.2 An Alternative Approach: Relational Accounts of Vulnerability in Research

In response to the previously detailed problems posed by guidelines on vulnerability, several authors have proposed alternative accounts of vulnerability in research, which take a more nuanced, dynamic, and relational view. In contrast to the labeling approach generally taken by guidelines, such accounts view vulnerabilities as something inessential and as a function of the relation between the specific research situation and the specific people in it, and the specific characteristics of both. On such a view, identifying what may make individuals vulnerable to exploitation necessarily involves taking into account and paying close attention to all the features of the situation and the people involved.

This line of thought is brought forth by Florencia Luna,[21] who—in opposition to the labeling approach taken by guidelines—offers what she terms a *layered* approach to vulnerability: rather than certain groups being vulnerable *per se*, they may in certain situations be "rendered vulnerable" by certain societal, cultural, and/or structural factors, and in this way "acquire certain overlapping

---

20 Carol Levine and others, "The Limitations of 'Vulnerability' as a Protection for Human Research Participants" (2004) 4(3) American Journal of Bioethics 44.

21 Luna, *supra* note 18.

layers of vulnerability."[22] Responding to the notion that women are sometimes considered vulnerable *simpliciter*, Luna illustrates her conception as follows:

> [b]eing a woman does not, in itself, imply that a person is vulnerable. A woman living in a country that does not recognize, or is intolerant of reproductive rights, acquires a layer of vulnerability ... [A]n educated and resourceful woman in that same country can overcome some of these consequences of the intolerance of reproductive rights; however, a poor woman living in a country intolerant of reproductive rights acquires another layer of vulnerability ... [A]n illiterate poor woman in a country intolerant of reproductive rights acquires still another layer.[23]

In this way, any individual stands to acquire a number of overlapping layers of vulnerability, depending on the situation and the circumstances. If one of the variables in the situation changes, she may no longer be considered vulnerable. As Luna notes, this richer, relational notion of vulnerability allows us to think of vulnerabilities as a function of economic, social, and cultural conditions and circumstances, and forces us to recognize that such conditions may invoke several different layers of vulnerability, operating on different levels and affecting different aspects of an individual's life.

In this vein, Kenneth Kipnis has offered a taxonomy of vulnerability,[24] identifying six dimensions on which individuals may be vulnerable with respect to their agreement to participate in research—or, in other words, how they may be open to exploitation in such research setups. Of Kipnis' six dimensions—cognitive, juridic, deferential, medical, allocational, and infrastructural[25]—I will address the juridic and deferential, as these (on my view) have been given least attention in the literature on exploitation in research and in research guidelines, and for this exact reason may be overlooked in the sort of research cases I intend to discuss later in this contribution.

Juridic vulnerability refers to the formal authority relationships that often characterize social structures, for example children who are under the authority of their parents, and certain Third World women who may be legally subject to their husbands. Such circumstances do not make individuals vulnerable *per se*—rather, the concern is that the juridic fact of their subordination to another may influence the validity of their consent, as it may merely be a reflection of the wishes of those in authority.[26]

Deferential vulnerability is closely related to juridic vulnerability, as it is to be viewed as the "deferential patterns [that are] subjective responses to

---

22 Luna *supra* note 18, at 122; 128–9.
23 *Ibid* at 128–9.
24 Kipnis, *supra* note 18.
25 *Ibid* at 4–9.
26 *Ibid* at 6.

certain others,"[27] often as a psychological effect of the objective features of the formal hierarchical context within which the individual finds herself. This can be exemplified by the (generally) intuitive obedience to a police officer, or the general tendency to comply with a doctor's orders. The vulnerability here is the individual's "readiness to accede to the perceived desires of certain others notwithstanding an inner reticence to do so."[28] Deferential vulnerability may emerge in any social relation, but is more easily identified where there is already a juridic relation to "flag" it. In settings where this is not the case, this type of vulnerability is much more easily overlooked, but no less powerful.

In what follows I explore these types of vulnerability in a concrete research case.

5 A CASE STUDY: EXPERIENCES FROM INTERVIEWS WITH BIOBANK DONORS FROM RURAL PAKISTAN

As an effort to explore the motivations, hopes and concerns of biobank donors in developing countries—a research aim set out by the *Global Genes, Local Concerns* project—our group of researchers from the University of Copenhagen conducted ten semi-structured interviews with families who had donated blood samples to a collaborating institute for genetic research, the National Institute for Biotechnology and Genetic Engineering (NIBGE) based in Faisalabad, Pakistan.[29]

Families had given samples for research on recessive genetic disorders, which several of a family's children were often afflicted with (due to the local practice of cousin marriages). To collect samples, NIBGE researchers traveled to donors' localities and encouraged them to donate samples, informing them that their samples will be used in research that may eventually lead to insights into the molecular basis of the particular disorder and how to control and prevent it in future generations. In the standard procedure researchers reach out to the elder of the family, honoring the local custom that the elder functions as the guardian of the family and has authority to make decisions on behalf of his family.

---

27 *Ibid.*

28 *Ibid.*

29 The study took place in April 2014, and is detailed in N.C.H. Kongsholm, J. Lassen, and P. Sandøe, "'I didn't have anything to decide, I wanted to help my kids': An Interview-based Study of Consent Procedures for Sampling Human Biological Material for Genetic Research in Rural Pakistan" (2018) AJOB Empirical Bioethics 113.

All families participating in interviews resided in villages in the rural areas surrounding the institute in Faisalabad, Pakistan. Family members had little to no education, and generally lived in conditions of severe poverty. All respondents expressed eagerness with respect to donating samples to NIBGE researchers, and all had consented either themselves or through a family elder.

5.1 Findings

As could perhaps be expected, donor respondents expressed that the hope of obtaining information about, or any sort of help regarding, the disorder afflicting their children was a highly motivating factor in their decision to participate in the research. Several families also linked this motivation to their poor financial situation, which meant that they had very few resources to care for their (often several) disabled children.

Apart from these motivations, however, a number of other themes recurred through the donors' responses—these were of particular interest as, after a closer look at the surrounding societal, religious, and cultural landscape, they appeared tightly linked to certain local customs and beliefs.

First, several respondents relayed that they were motivated to donate samples because the elder of their family—typically their husband, their father, or an older uncle—had decided that their family should give samples to research. This should be viewed in light of the social landscape in Pakistan: the country is often described as characterized by a highly hierarchical structure both in public matters and extending into the private sphere. Many commentators have described Pakistani society as family-centered, meaning that the family, not the individual, is the held to be the fundamental unit of society and decisionmaking.[30] This is reflected in housing arrangements: especially in rural areas, it is not uncommon to see several generation of the same family living under one roof, pooling their resources.[31] Respondents often mentioned power structures being clearly defined along lines of sex and age and wisdom being attributed to age, commanding respect, loyalty, and obedience.[32] Similarly, families were often described as accrediting decisionmaking authority to the elder of the family, as they are considered to hold sufficient wisdom to know what is in the best interest of members of the family and the family as

---

30 Farhat Moazam, "Families, Patients and Physicians in Medical Decisionmaking: A Pakistani Perspective" (2000) 30(6) Hastings Center Report 28.

31 Farhat Moazam and Aamir M. Jafarey, "Pakistan and Biomedical Ethics: Report from a Muslim Country" (2005) 14 Cambridge Quarterly of Healthcare Ethics 249.

32 See for example Riffat M. Zaman, "Psychotherapy in the Third World: Some Impressions from Pakistan," in Uwe P. Gielen and others (eds), *Psychology in International Perspective* (Swets & Zeitlinger 1992).

a whole.[33] It is thus not uncommon—and is also legally legitimate—for women to have decisions made for them by their husband or father, or for family elders to decide on behalf of their entire family.

Second, many respondents explained that the fact that the NIBGE researchers were associated with the medical profession, and in virtue of this were held to know the best course of action for the family, was conducive to their decision to donate samples. It is interesting to view this sort of motivation in relation to the religious landscape in Pakistan, and in particular the religious role accorded to medical professionals. In total, 95 percent of Pakistani citizens are Muslims, and the country's sociocultural life and values are greatly shaped and informed by Islam. Religious belief plays a central role in the life of both men and women from all social strata, and is a major influence on all public and private activities. According to Islamic belief, the profession of doctor is regarded as highly respectable, imbuing representatives with elevated, almost divine, insights and authority. As Farhat Moazam writes:

> Reverence and respect toward physicians is due not only to their knowledge and scientific expertise but also to the historical position accorded to the art and science of medicine in Islam. The privileged position of physicians is derived through a historical understanding of the healer as an instrument of divine mercy.[34]

This places an extraordinary authority on doctors: "The 'doctor sahib'[35] ... is expected to direct rather than just facilitate medical management ... Interaction with a physician thus takes the form of recourse to an authority figure and not merely a consultation with a medical expert."[36]

For these reasons, representatives from the medical profession, including both medical doctors and researchers, are held to possess greater insight and authority than individuals not associated with this field, and their directives or recommendations are generally—and, indeed, often automatically—followed.

5.2 Vulnerabilities in the Pakistan Case

Returning to the lessons learned from the previous section on the nature of vulnerability, we may argue that the donor motivations just accounted for can be interpreted as expressions of vulnerability in this sense.

---

[33] Farhat Moazam, *Bioethics and Organ Transplantation in a Muslim Society* (Indiana University Press 2006).

[34] *Supra* note 29, at 31.

[35] From Arabic; meaning "lord."

[36] *Supra* note 29, at 28–31.

In cases where daughters and wives of families had consented to giving samples motivated by their father, husband, or elder's decision that the family should donate, there is reason to question whether these women were, recalling Kipnis' terminology, *juridically* vulnerable, in the sense that, due to ingrained social and hierarchical norms of elder and male authority, they may not consider themselves to have the option to *not* follow decisions set forth by the family head. Note that this is *not* equal to claiming that the women are forced or coerced into donation by their husbands or fathers; rather, it may be described as a social reflex to comply with the elder's directives. Kamuya and colleagues have described this tendency in the context of biomedical research in Kenya:[37]

> [M]any mothers consider decision making by the child's father as the social norm … the general expectation that decisions about a child's participation in research are made by the father means that the mother is simply following her social obligations. In our experience, a mother in this situation might be puzzled by a question of whether or not she voluntarily decided to ask the child's father to make the decision on her behalf.

Tindana and colleagues have found similar tendencies among community members in northern Ghana, stating: "[W]hen leaders give permission for a study to go forward, community members may view this as an endorsement of the study, which may in turn significantly influence research participation."[38] In our case, a vulnerability may arise if the wife or daughter of a family does not want to donate her blood sample (as per her elder's decision), but, due to social constraints, does not view it as an option to go against the wishes of her elder. As consent in this case is obtained by approaching the elder, this means that his decision regarding family participation is established first, and is subsequently relayed to his family. In this way the wife or daughter is subject to his authority regarding participation, and is thus—to employ the terms of Luna and Kipnis—rendered juridically vulnerable.

A similar tendency can be observed in cases where donors decided to participate because of the researchers' status as part of the medical profession. As stated previously, there is in this local area a tendency to ascribe almost divine authority to medical professionals, as people whose orders are to be followed. This is primarily the case with physicians, but there is reason to believe this

---

37 D. Kamuya, V. Marsh, and S. Molyneux, 'What We Learned about Voluntariness and Consent: Incorporating 'Background Situations' and Understanding into Analyses' (2011) American Journal of Bioethics, 11(8), 31–3, at 31–2.

38 P.O. Tindana, N. Kass, and P. Akweongo, "The Informed Consent Process in a Rural African Setting: A Case Study of the Kassena-Nankana District of Northern Ghana" (2006) IRB: Ethics & Human Research, 28(3), 1–6, at 6.

authority can spill over to medical researchers—especially in settings where, due to low education levels, individuals may not be familiar with the differences between medical practice on one hand and research on the other (this tendency has been observed in similar studies in similar settings).[39] In this way, it may be the case that donors are *deferentially vulnerable*: in a setting where obedience to the directives of medical professionals is the social norm, it is possibly the case that donors do not consider the option to donate as a genuine choice, but rather as following the recommendation of an authority figure. As above, it would not be fitting to describe this situation as coercive; rather, it is likely that due to ingrained deferential patterns to certain people—in this case, the medical researchers—donors simply do not regard saying no as an option, and may even embrace the "recommendation" to donate with joy and gratitude.

I have discussed examples of how the donors in our case, due to local conventions and customs, may be vulnerable. I have not, however, argued that they are necessarily being exploited because of this fact. However, it should be fairly clear how the present vulnerabilities indeed make exploitation a possibility in this setting. This could for example be the case if researchers approach the family elder for consent to donation not merely because it is the social custom, but because they know that other family members will likely comply with his decision; or if researchers take advantage of a donor's inclination to obey medical professionals by emphasizing their affiliation with this profession (and downplaying the fact that donors are not likely to receive any direct benefit from their participation, as they would (in the form of a cure or the like) in the case of consultation with a physician). These would be cases of deliberate exploitation by local researchers; however, it can also be stipulated that the same vulnerabilities can lead to unknowing exploitation by foreign researchers. This could be the case if consent given by the women of a family through the elder is taken to equal their full voluntary consent to participation even though this may not be the case (as we saw above), or in cases where the particular cultural norm of behaving in a certain deferential way toward medical professionals is mistaken for simple friendliness and cooperation.

---

[39] See for example Kamuya and others, *supra* note 36; Caroline Gikonyo and others, "Taking Social Relationships Seriously: Lessons Learned from the Informed Consent Practices of a Vaccine Trial on the Kenyan Coast" (2008) 67(5) Social Science & Medicine 708.

6 IMPLICATIONS FOR RESEARCH POLICY: GUIDELINES AND HARMONIZATION

The lessons to be learned from the preceding discussion are as follows: vulnerabilities to exploitation in research are subtle and highly contextually and culturally dependent, and what makes individuals vulnerable to certain other individuals and in certain situations may vary greatly from one cultural setting to another. Furthermore, such vulnerabilities only become fully visible if one is sufficiently familiar with the background norms and customs of the cultural setting at hand, and how these give rise to vulnerabilities.

The upshot is that if researchers operating in a particular cultural setting (especially one that is foreign from their own) are not acutely aware of its particular vulnerability-generating factors, these become sources of potential exploitation, at risk of being overlooked.

This has implications for the development of protocols, policies, and guidelines for research in developing countries. As we have seen, the present focus on consent as an overarching ethical seal of approval is not adequate, for two reasons: (1) even consensual interactions can be exploitative, and (2) superficially valid formal consent can be, and often is, compromised by local, cultural factors that are not apparent without adequate familiarity with the particular research settings in which they play out. Kipnis and Luna each put forth suggestions as to how accommodating these shortcomings in practice requires a two-pronged effort.[40]

6.1 Thorough Identification and Analysis of All Present Vulnerabilities in the Given Research Context

The first prong requires not only consideration of the potential risks and benefits of the research and the attributes of the population, but also a competent analysis of the broader social and cultural context, and scrutinizing the specific research protocol in this light. This includes being acutely aware of any elements of the research protocol and research environment that may affect the particular subjects in a certain way, such as certain types of verbal and symbolic communication that may prompt a certain response from certain subjects.

---

[40] See also Hurst, *supra* note 18 at 198–9 for an example of application of this approach by an IRB.

6.2 Developing Appropriate Measures to Accommodate Each Type of Vulnerability

This step requires the development of an appropriate response to each vulnerability identified through the foregoing analysis. Here, it is important to note that each may require a different mechanism for its neutralization with respect to exploitation, and may address different levels of the research. As examples, Kipnis notes that mechanisms that may accommodate juridic vulnerability may include devising a consent procedure that will adequately insulate the individual from the hierarchical system and its juridic powers, and mechanisms that accommodate deferential vulnerability may include paying attention to the conversational setting and devising a procedure that eliminates as much as possible of the social pressure an individual may feel she is under (even if this is not the case).

In a biobanking context, these considerations in turn have implications for the current efforts to harmonize guidelines and standards, particularly with regard to consent. It should be clear from the preceding discussion that complete harmonization and standardization with respect to consent is—without qualification—not a desirable goal, as it risks glossing over local factors that compromise consent and thus paving the way for exploitative research interactions. It is worthwhile here to invoke Kipnis' point that "in the minds of many investigators the paradigmatic research subject remains more or less a mature, respectable, moderately well-educated, clear-thinking, literate, self-supporting U.S. citizen in good standing";[41] this indicates that where one type of consent may be adequate and appropriate for such individuals, it may not be so for individuals who do not fulfill these criteria. Hence there is a need to supplement harmonization efforts with attention to local factors that may compromise consent in developing countries (for example, in the ways suggested by Kipnis and Luna).

---

41 Kipnis, *supra* note 19 at 1.

10. Biobanking and the consent problem

Timothy Caulfield and Blake Murdoch[1]

1 INTRODUCTION

Driven by advances in genetics, information technology, and cell-line research, interest in the collection and analysis of human biological material continues to intensify. Over the past few decades, there has been a proliferation of biobanks—both large and small[2]—that link tissue and genetic information to a host of other forms of health and personal data. Indeed, biobanking and related research methods have been characterized as an essential and potentially "revolutionizing" approach to biomedical research.[3] In 2009, for example, a *Time Magazine* cover story framed biobanking as a "world changing" idea.[4] Since then, the enthusiasm has not diminished. Governments and industries throughout the world have invested heavily in biobanking.[5] This is perhaps best exemplified by former President Barack Obama's championing

---

[1] The authors would like to thank Marcello Tonelli and Robyn Hyde-Lay for their input and suggestions. Finally, the authors would like to thank all participants in the Health Law Institute workshop titled "Emerging Research Methods and the Consent Challenge," held September 25–27, 2016, in Banff, Alberta, Canada. This chapter was first published in PLOS Biology on July 25, 2017.

[2] Gail E. Henderson et al, "Characterizing biobank organizations in the US: results from a national survey" (2013) Genome Medicine 5(1) 3.

[3] Futurism, "U.S. to collect genetic data for biobank, what this means for you" (*Futurism*, February 16, 2015). https://futurism.com/u-s-collect-genetic-data-biobank-means/ accessed October 4, 2017.

[4] Alice Park, "10 ideas changing the world right now: biobanks" (*Time*, March 12, 2009) http://content.time.com/time/specials/packages/article/0,28804,1884779_1884782_1884766,00.html accessed October 4, 2017.

[5] Jimmie Vaught et al, "Biobankonomics: developing a sustainable business model approach for the formation of a human tissue biobank" (2011) Journal of the National Cancer Institute (42) 24.

of the Precision Medicine Initiative,[6] which, among other things, included the creation of a large national biobank.[7]

In addition to the excitement flowing from the scientific and medical potential of biobanking, there has been a great deal of controversy. Much of the conflict has centered on issues of consent and the control of tissue samples. Because of the large number of participants (UK Biobank, for example, recruited 500,000 individuals between 2006 and 2010[8]), the involvement of multiple researchers, and the long-term nature of the initiatives, the use of traditional models of consent is considered impractical, if not impossible. As such, biobanks have adopted consent strategies that deviate from traditional legal norms, most often involving the use of some form of broad or open consent.[9] But despite the mass utilization of these approaches, there remains no consensus as to their legal and ethical appropriateness. A 2012 analysis of relevant literature, research ethics policies, and public perception data found that, aside from within the biobanking research community, there is no consensus on the consent issue.[10] And there is little reason to think that a consensus will coalesce in the future, despite attempts by some jurisdictions, including the US and the European Union, to craft relevant policy.[11]

We live in a fascinating and potentially precarious time. On the one hand, growth, investment, and excitement surrounding biobanking continue to escalate. Across the globe, millions of individuals have been recruited to participate in these complex, expensive research platforms, and the potential scientific and health benefits are undoubtedly real. On the other hand, there remains a deep lack of clarity around basic legal and ethical principles. The public is supportive, but that support appears tentative and conditional on the

---

6 Sharon F Terry, "Obama's precision medicine initiative" (2015) Genetic Testing and Molecular Biomarkers 19(3) 113.

7 National Institutes of Health, "NIH funds biobank to support Precision Medicine Initiative Cohort Program" (*National Institutes of Health*, May 26, 2016) www.nih.gov/news-events/news-releases/nih-funds-biobank-support-precision-medicine-initiative-cohort-program accessed October 4, 2017.

8 UK Biobank, "About UK Biobank" (*UK Biobank*, 2016) www.ukbiobank.ac.uk/about-biobank-uk/ accessed October 4, 2017.

9 Theresa Edwards et al, "Biobanks containing clinical specimens: defining characteristics, policies, and practices" (2014) Clinical Biochemistry 47(4) 245.

10 Zubin Master et al, "Biobanks, consent and claims of consensus" (2012) Nature Methods 9(9) 885.

11 Dara Hallinan and Michael Friedewald, "Open consent, biobanking and data protection law: can open consent be 'informed' under the forthcoming data protection regulation?" (2015) Life Sciences, Society and Policy 11(1) 1; Federal Register, "Federal Policy for the Protection of Human Subjects" (*Federal Register*, January 19, 2017) www.federalregister.gov/documents/2017/01/19/2017-01058/federal-policy-for-the-protection-of-human-subjects accessed October 4, 2017.

maintenance of trust. The bottom line: the international research community has built a massive and diverse research infrastructure on a foundation that has the potential, however slight, to collapse, in part or altogether. Those most involved in the research—that is, those involved with the collection of samples and the establishment and administration of biobanks—appear to be operating under the belief that the issues associated with the law and public opinion are either settled or manageable within existing frameworks. Here, we seek to highlight how wrong such assumptions are. What is needed is real policy reform. We believe this would benefit from more explicit recognition of the vast disconnect between the current practices and the realities of the law, research ethics, and public perceptions.

Of course, concerns about consent and ownership are hardly new. On the contrary, biobanking has gained support and flourished despite the ongoing and frequently articulated apprehension surrounding issues like consent and ownership. However, there are emerging social trends and technological developments—several of which we review in this chapter—that have heightened the need for increased clarity in the context of consent policy.

2 UNSETTLED LAW

One would expect that with so much activity and such broadbased investment, the law relevant to biobanking would be relatively settled, especially for something as fundamental to the research ethics process. In fact, there remains a great deal of uncertainty regarding, inter alia, the ownership of samples,[12] as well as what type of consent is legally appropriate. A recent review of Canadian law by a team at McGill University, for example, concluded that while broad consent is ubiquitous, it "does not seem to fulfill legal and ethical informational requirements."[13] It has been suggested that emerging regulations in Europe dictate that "consent must be 'specific and informed'."[14] A 2017

---

[12] R. Alta Charo, "Body of research—ownership and use of human tissue" (2006) New England Journal of Medicine 355(15) 1517; Timothy Caulfield and Amy L. McGuire, "Policy uncertainty, sequencing, and cell lines" (2013) G3: Genes, Genomes, Genetics 3(8) 1205; Jennifer K. Wagner, "Property rights and the human body" (*Genomics Law Report*, June 11, 2014) www.genomicslawreport.com/index.php/2014/06/11/property-rights-and-the-human-body/ accessed October 4, 2017.

[13] Clarissa Allen, Yann Joly, and Palmira Granados Moreno, "Data sharing, biobanks and informed consent: a research paradox" (2013) McGill JL & Health 7 85.

[14] Council Regulation (EU) 2016/679 of the European Parliament and of the Council on the protection of natural persons with regard to the processing of personal data and on the free movement of such data, and repealing Directive 95/46/EC (General Data Protection Regulation) [2016] OJ L 119; Hallinan and Friedewald, *supra* at note 11.

revision of the US Common Rule, the country's national research ethics guideline, explicitly endorses the use of broad consent in specific situations,[15] but this policy change does not resolve debate about whether such an approach is appropriate; as noted by a commentator, "potential subjects cannot be informed of the specific risks and benefits of research because the biobanks do not know what those risks or benefits may be."[16] It has been suggested, for example, that the new regulations—which disappointed patient advocacy groups[17]—reflect the power of the research institutions' lobby more than any conceptually and legally coherent policy change.[18]

Here, we offer no speculation about what the law ought to be (a topic one of us has covered elsewhere[19]). Our key point is straightforward: despite decades of debate and a huge amount of public and private investment in biobanking, there is still a great deal of ambiguity and uncertainty regarding the issues associated with participant control of specimens and health information.[20] Yet despite this reality, biobanks throughout the world have forged ahead using various alternatives to traditional models of consent and governance structures.[21] Given that this research method seems unlikely to go away any time soon, it is worthwhile to reflect on the building tension between the enthusiasm for biobanks and the need for a robust and conceptually consistent framework for consent.

---

15 Holly F. Lynch, "Final Common Rule revisions just published" (*Harvard Law: Bill of Health*, January 18, 2017) http://blogs.harvard.edu/billofhealth/2017/01/18/final-common-rule-revisions-just-published/ accessed October 4, 2017; James G. Hodge and Lawrence O. Gostin, "Revamping the US Federal Common Rule" (*JAMA*, February 22, 2017) http://jamanetwork.com/journals/jama/fullarticle/2606525 accessed October 4, 2017; Joshua D Smith et al, "Immortal life of the Common Rule: ethics, consent, and the future of cancer research" (2017) Journal of Clinical Oncology 35(17) 1879.

16 Hank T. Greely, "The uneasy ethical and legal underpinnings of large-scale genomic biobanks" (2007) Annu. Rev. Genomics Hum. Genet. 8 343.

17 Sara Reardon, "Controversial patient-consent proposal left out of research-ethics reforms" (2017) Nature 541(7638) 449.

18 Jocelyn Kaiser, "U.S. abandons controversial consent proposal on using human research samples" (*Science*, January 18, 2017) www.sciencemag.org/news/2017/01/update-us-abandons-controversial-consent-proposal-using-human-research-samples accessed October 4, 2017.

19 Timothy Caulfield et al, "Research ethics recommendations for whole-genome research: consensus statement" (2008) PLoS Biol 6(3) e73.

20 Timothy Caulfield et al, "A review of the key issues associated with the commercialization of biobanks" (2014) Journal of Law and the Biosciences 1(1) 94.

21 Master, *supra* at note 10; Greely, *supra* at note 16; Palmira Granados Moreno and Jann Yoly, "Informed consent in international normative texts and biobanking policies: seeking the boundaries of broad consent" (2016) Medical Law International 15(4) 216.

3 PUBLIC PERCEPTION

A great deal of public perception research has been done on the issues associated with biobanking. While there are few uniform messages that flow from this body of data, one thing can be said with certainty: there is no consensus on how to handle consent. For example, a 2016 study in the US found that "nearly 44% of our nationally representative sample found blanket consent unacceptable and 38% felt it was, in fact, the worst in a range of consent policy options."[22] Other research has also found that a significant portion of the public is not keen on the use of broad, open, or blanket consent if other options are available.[23] Indeed, a 2015 systematic review of the available data concluded that "the most notable finding is that many people do not favor broad consent for either research itself or for research and subsequent wide data sharing."[24] Yes, research has found that many in the public find broad consent acceptable,[25] but this does not mean that it is the preference or that the approach is without controversy. Naturally, a broad or general consent approach is clearly preferred by biobank researchers.[26] But, as reported in a 2015 survey, the scientific community also "does not believe there to be a consensus on consent type."[27]

When it comes to the public's perception of consent in the context of biobanks, the most that can be said is that it remains unsettled,[28] and there seems little reason to think a unified position will emerge. In fact, as we will see,

22 Raymond G. De Vries et al, "Understanding the public's reservations about broad consent and study-by-study consent for donations to a biobank: results of a national survey" (2016) PloS ONE 11(7) e0159113.

23 Juli Murphy et al, "Public perspectives on informed consent for biobanking" (2009) American Journal of Public Health 99(12) 2128; Christian M. Simon et al, "Active choice but not too active: public perspectives on biobank consent models" (2011) Genetics in Medicine 13(9) 821.

24 Garrison Nanibaa'A et al, "A systematic literature review of individuals' perspectives on broad consent and data sharing in the United States" (2015) Genetics in Medicine 18(7) 663.

25 Saskia C. Sanderson et al, "Public attitudes toward consent and data sharing in biobank research: a large multi-site experimental survey in the US" (2017) American Journal of Human Genetics 100(3) 414.

26 Zubin Master, Lisa Campo-Engelstein, and Timothy Caulfield, "Scientists' perspectives on consent in the context of biobanking research" (2015) European Journal of Human Genetics 23(5) 569.

27 Zubin Master, "The US National Biobank and (no) consensus on informed consent" (2015) American Journal of Bioethics 15(9) 63.

28 Murphy, *supra* at note 23; Simon, *supra* at note 23; Garrison, *supra* at note 24; Master, *supra* at note 27.

there are several social trends that suggest the environment may become even more confused.

3.1 Perceived Rights of Control

The growing interest in the concept of biorights could be particularly disruptive. This refers to the idea that research participants have an ongoing right to control their research samples, to benefit directly from the research, and/or to be financially compensated for their contribution.[29] While the existing public perception research suggests that this view is likely held by only a minority—most in the public remain willing to contribute to biobanks and similar initiatives without these terms[30]—it would not take many individuals supportive of the idea of biorights to complicate the consent process and the concomitant public discourse, as we have seen with a number of related controversies.[31] For example, the infamous case of Henrietta Lacks, who was the source of the HeLa cell line, continues to stir controversy. Members of the family are now seeking financial compensation for the cells,[32] and the story of Henrietta Lacks was made into a movie starring Oprah Winfrey.[33]

Ironically, the growing interest in a perceived right of control may be due, in part, to exaggerated claims of benefit flowing from both the research community and the popular press. A 2014 study found that many of the representations of biobanks in the popular press are "hyped"—that is, they contain exaggeration of both the near future benefits and minimization of risks and limitations.[34]

---

29 Beth Daley and Ellen Cranley, "Biorights' rise: donors demand control of their samples" (*Boston Globe*, October 10, 2016) www.bostonglobe.com/metro/2016/10/09/the-rise-biorights-donors-are-demanding-control-and-sometimes-cash-exchange-for-genetic-samples/jCbaQ2E5t6c0Qs1kcITMRM/story.html accessed October 4, 2017; Brenda Lau, "Patients are more aware about their 'biorights' and demand to be compensated" (*MIMS News*, October 22, 2016) http://today.mims.com/topic/patients-are-more-aware-about-their-biorights-and-demand-to-be-compensated accessed October 4, 2017.

30 Janet L. Cunningham et al, "No regrets: young adult patients in psychiatry report positive reactions to biobank participation" (2017) BMC Psychiatry 17(1) 21; Sanderson, *supra* at note 25.

31 Rebecca Skloot, *The Immortal Life of Henrietta Lacks* (Broadway Paperbacks 2010).

32 Erin Blakemore, "New claims prove the Henrietta Lacks controversy is far from over" (*Smithsonian*, February 15, 2017) www.smithsonianmag.com/smart-news/claims-henrietta-lacks-controversy-far-from-over-180962185/ accessed October 4, 2017.

33 Tina Jordan, "See the first photos of Oprah Winfrey in HBO's Henrietta Lacks movie" (*Entertainment Weekly*, January 22, 2016) http://ew.com/tv/2016/12/22/oprah-winfrey-hbo-henrietta-lacks-movie/ accessed October 4, 2017.

34 Ubaka Ogbogu et al, "Newspaper coverage of biobanks" (2014) PeerJ e500.

And, of course, genetic research more broadly has been the subject of a great deal of overpromise.[35] The same can be said for stem cell research.[36]

All of this positive coverage—accurate or not—may contribute to the perception that genetic information and human cells are especially sensitive, valuable, and worthy of unique protections and individual control. Headlines such as "Cell Line Development Market worth $3.96 Billion by 2019,"[37] "Human Biobanking Ownership—Market to Witness a Value of $37.1 Billion by 2020,"[38] and "Could We One Day Make Babies from Only Skin Cells?"[39] add to the impression that there may be many reasons to retain a strong right of control over donated research samples. A 2016 American news item—which also offered yet another estimation of value, suggesting that biological samples collected by researchers might generate "$23 billion in revenue by 2018"—quoted a research participant who demanded and received payment for her biological materials.[40] The research participant suggested "there has been an over-assumption and a gross expectation of patient altruism."[41] This kind of reaction fits well with a prediction made by Stanford law professor Hank Greely in 2010: "As more and more people find out what can be done—or is being done—with their health information, their family histories, and their DNA, the pressure for change should grow."[42]

When thinking about the forces that drive social controversy, it does not matter whether these views regarding biorights and the value of genetic information are justified. In fact, there are reasons to dispute the idea that genetic information is uniquely valuable,[43] and, as one of us has argued elsewhere,

---

35 Eliot Marshall, "Waiting for the revolution" (2011) Science 331(6017) 526.

36 Timothy Caulfield et al, "Confronting stem cell hype" (2016) Science 352(6287) 776.

37 PR Newswire, "Cell line development market worth $3.96 billion by 2019" (*PR Newswire*, September 22,2014) www.prnewswire.com/news-releases/cell-line-development-market-worth-396-billion-by-2019-275982961.html accessed October 4, 2017.

38 Global Market Watch, "Human biobanking ownership market to witness a value of US$37.1 billion by 2020" (*MedGadget*, February 24, 2017) www.medgadget.com/2017/02/human-biobanking-ownership-market-to-witness-a-value-of-us37-1-billion-by-2020.html accessed October 4, 2017.

39 Kelly Murray, "Could we one day make babies from only skin cells?" (*CNN*, February 9, 2017) www.cnn.com/2017/02/09/health/embryo-skin-cell-ivg/index.html accessed 4 October 2017.

40 Daley, *supra* at note 29.

41 *Ibid.*

42 Hank T. Greely, "To the barricades!" (2010) Am. J. Bioeth. 10(9) 1.

43 James P. Evans and Wylie Burke, "Genetic exceptionalism: too much of a good thing?" (2008) Genetics in Medicine 10(7) 500.

public perception should not necessarily drive policy.[44] Yet it often does.[45] Public perception can also drive public debate and can serve as a barometer of future social controversy.

3.2 Public Trust and Commercialization

Surveys have consistently found that the public places a great deal of trust in researchers and research institutions. However, public perception studies have also found that trust can be easily lost. Any involvement with industry, for example, erodes public confidence in the biobanking enterprise.[46] A 2012 survey of 1,201 people in Alberta, Canada found that 45.1 percent had a "great deal" of trust in university-funded researchers, but only 19.5 percent felt the same way about university researchers who received funding from industry.[47] The number drops to 6 percent for biobanking research done by industry.[48] Commercial interests seem to have the same impact on how the public perceives the acceptability of various forms of consent. A 2016 study found that a majority of individuals (68 percent) were willing to provide a blanket consent, but that drops to 55 percent if their specimens might be used "to develop patents and earn profits for commercial companies."[49]

One of the key goals of biobanks is to produce technologies and drugs that will be used in the healthcare context. Industry will inevitably be involved, particularly as the work gets closer to clinical application. As argued in a 2015 commentary, "ultimately, the success of future biobanks will rely greatly on the success of public-private partnerships."[50] In addition, high maintenance

---

44 Timothy Caulfield, "Biobanks and blanket consent: the proper place of the public good and public perception rationales" (2007) King's Law Journal 18(2) 209.

45 Paul Burstein, "The impact of public opinion on public policy: a review and an agenda" (2003) Political Research Quarterly 56(1) 29.

46 Christine R. Critchley and Dianne Nicol, "Understanding the impact of commercialization on public support for scientific research: is it about the funding source or the organization conducting research?" (2009) Public Understanding of Science 20(3) 347.

47 Timothy Caulfield, Christen Rachul, and Erin Nelson, "Biobanking, consent, and control: a survey of Albertans on key research ethics issues" (2012) Biopreservation and Biobanking 10(5) 433.

48 *Ibid.*

49 Tom Tomlinson et al, "Moral concerns and the willingness to donate to a research biobank" (2015) JAMA 313(4) 417; Raymond G. De Vries and Tom Tomlinson, "Americans want a say in what happens to their donated blood and tissue in biobanks" (*The Conversation*, July 13, 2016) http://theconversation.com/americans-want-a-say-in-what-happens-to-their-donated-blood-and-tissue-in-biobanks-60681 accessed October 4, 2017.

50 Stella B. Somiari and Richard I. Somiari, "The future of biobanking: a conceptual look at how biobanks can respond to the growing human biospecimen needs of

costs mean that many biobanks must turn to industry for funding support, which some fear could affect the independence and integrity of research, or result in changes in the use of data or samples that are inconsistent with existing consents.[51] The inevitability of increased industry involvement creates the potential to erode public trust and to intensify the challenges associated with consent and sample ownership.[52]

This push to commercialize can also have a direct impact on the consent process, heightening the need for specific consent or, at least, specific disclosures about commercialization and industry involvement. The importance of informing participants about commercialization and industry activities has been recognized as a key part of the consent process: this information is something participants want to know.[53] Indeed, studies have found that "information about sponsoring of biobank research by pharmaceutical industry was associated negatively with a preference for broad consent."[54] It is also worth noting that Article 26 of the European Union's Directive 98/44/EC on the legal protection of biotechnological inventions states that before issuing a patent, it should be established that free and informed consent has been obtained from individuals who have contributed the relevant biological material.[55] In sum, increasing commercialization pressure seems likely to intensify the already complex consent and control issues associated with biobanks.

3.3 Privacy and Discrimination Concerns and Controversies

Public interest in maintaining control over biobanked information and samples may be heightened by highly publicized instances of the mishandling of private health information. Data breaches involving confidential medical

---

researchers," in *Biobanking in the 21st Century* (Springer International Publishing 2015).

51 Caulfield, *supra* at note 20.

52 *Ibid.*

53 *Ibid*; Canadian Institutes of Health Research, Natural Sciences and Engineering Research Council of Canada, and Social Sciences and Humanities Research Council of Canada, "Tri-Council Policy Statement: ethical conduct for research involving humans" (2014) Article 3.2 www.pre.ethics.gc.ca/pdf/eng/tcps2-2014/TCPS_2_FINAL_Web.pdf accessed October 4, 2017.

54 Flavio D'Abramo, Jan Schildmann, and Jochen Vollmann, "Research participants' perceptions and views on consent for biobank research: a review of empirical data and ethical analysis" (2015) BMC Medical Ethics 16(1) 60.

55 Council Directive (EC) 98/44 on the legal protection of biotechnological inventions [1998] OJ L213:13–21.

information are increasing in number.[56] For example, in Georgia, the Emory Clinic experienced unauthorized data access in March 2017 that resulted in 79,930 individuals' personal information being compromised.[57] Similarly, Alberta Health Services recently experienced a privacy breach involving the confidential health information of 12,848 individuals.[58] These stories fuel growing public concern over the protection of personal health information.[59] Survey data indicate that members of the public worry about loss of control over their data, unauthorized use of it, and potential associated harms, including the possibility that health information could be wrongly used to inform discriminatory corporate, institutional, or public policies.[60]

Several issues compound these privacy concerns in the context of biobanking. First, there is survey evidence indicating that biobank participants are sometimes not fully aware of the confidentiality risks inherent to their participation.[61] Given the research finding that privacy issues are important to biobank participants,[62] any mishandling of the disclosure process by biobanks could have a severe adverse impact on public trust. Second, rightly or not, the public believes genetic information is particularly sensitive[63]—a perception

---

56 Alexandra W. Pecci, "Healthcare data breaches up 40% since 2015" (*HealthLeaders Media*, February 26, 2017) www.medpagetoday.com/practicemanagement/informationtechnology/63410 accessed October 4, 2017.

57 Thinkstock, "79K patients affected by Emory Healthcare data breach" (*HealthITSecurity*, March 2, 2017) http://healthitsecurity.com/news/79k-patients-affected-by-emory-healthcare-data-breach accessed October 4, 2017.

58 CBC News, "Alberta Health Services notifies almost 13,000 patients of privacy breach" (*CBC News*, September 26, 2016) www.cbc.ca/news/canada/edmonton/alberta-health-services-notifies-almost-13-000-patients-of-privacy-breach-1.3779296 accessed October 4, 2017.

59 Chrysanthi Papoutsi et al, "Patient and public views about the security and privacy of Electronic Health Records (EHRs) in the UK: results from a mixed methods study" (2015) BMC Medical Informatics and Decision Making 15(1) 86; Phoenix Strategic Perspectives Inc, "2016 survey of Canadians on privacy" (*Office of the Privacy Commissioner of Canada*, December 2016) www.priv.gc.ca/en/opc-actions-and-decisions/research/explore-privacy-research/2016/por_2016_12/?WT.ac=dpd-17&WT.ad=dpd-17 accessed October 4, 2017.

60 Phoenix, *supra* at note 59; Mhairi Aitken et al, "Public responses to the sharing and linkage of health data for research purposes: a systematic review and thematic synthesis of qualitative studies" (2016) BMC Medical Ethics 17(1) 73.

61 Laura A. Siminoff et al, "Confidentiality in biobanking research: a comparison of donor and nondonor families' understanding of risks" (2017) Genetic Testing and Molecular Biomarkers 21(3) 171.

62 David J. Kaufman et al, "Public opinion about the importance of privacy in biobank research" (2009) American Journal of Human Genetics 85(5) 643.

63 Rene Almeling and Shana K. Gadarian, "Public opinion on policy issues in genetics and genomics" (2013) Genetics in Medicine 16(6) 491; Susanne B. Haga et

likely fueled by overly optimistic media portrayals and cautionary stories about the potential for genetic discrimination.[64] As a result, any privacy issue that engages genetics is likely to heighten the public's desire to maintain control. Finally, the ability to pull information from cells, tissue, and genetic material has advanced rapidly over the past few decades. Indeed, the sequencing of entire genomes has become increasingly inexpensive and routine. This digitization of tissue, cells, and genetic data means that the line between health information (or health records) and tissue has largely disappeared.[65]

3.4 Technology and Consent Opportunities

The core justification for a move away from specific, case-by-case consent is that it is inefficient and expensive. But as information technology moves forward, that justification is weakened.[66] New tools are being developed to easily and inexpensively allow donors and participants to remain in constant contact with researchers and biobanks.[67] These tools can be used to allow participants whose health information and biological materials were collected by biobanks before the age of majority to reconsent or withdraw consent as adults, a practice which some suggest is an ethical necessity.[68] Indeed, the use of elec-

---

al, "Public knowledge of and attitudes toward genetics and genetic testing" (2013) Genetic Testing and Molecular Biomarkers 17(4) 327; Alicia A. Parkman et al, "Public awareness of genetic nondiscrimination laws in four states and perceived importance of life insurance protections" (2015) Journal of Genetic Counseling 24(3) 512; Annet Wauters and Ine Van Hoyweghen, "Global trends on fears and concerns of genetic discrimination: a systematic literature review" (2016) Journal of Human Genetics 61(4) 275.

[64] Julia Belluz, "A new bill would allow employers to see your genetic information—unless you pay a fine" (*Vox*, March 13, 2017) www.vox.com/policy-and-politics/2017/3/13/14907250/hr1313-bill-genetic-information accessed October 4, 2017.

[65] Ubaka Ogbogu et al, "Policy recommendations for addressing privacy challenges associated with cell-based research and interventions" (2014) BMC Medical Ethics 15(1) 7; Jennifer Kulynych and Hank T. Greely, "Clinical genomics, big data, and electronic medical records: reconciling patient rights with research when privacy and science collide" (2017) Journal of Law and the Biosciences 1 1.

[66] Jane Kaye et al, "Dynamic consent: a patient interface for twenty-first century research networks" (2015) European Journal of Human Genetics 23(2) 141; Isabelle Budin-Ljøsne et al, "Dynamic consent: a potential solution to some of the challenges of modern biomedical research" (2017) BMC Medical Ethics 18(1) 4.

[67] Haridimos Kondylakis et al, "Donor's support tool: Enabling informed secondary use of patient's biomaterial and personal data" (2017) International Journal of Medical Informatics 97 282.

[68] Noor A. Giesbertz, Annelien L. Bredenoord, and Johannes J. van Delden, "When children become adults: should biobanks re-contact?" (2016) PLoS Med 13(2) e1001959.

tronic consent systems is "a feasible and potentially game-changing strategy" for large research studies that depend on patient recruitment.[69] It is possible that the public may increasingly view broad consent as inappropriate in light of the ability to use technological tools to efficiently provide for reconsent and related forms of specific consent.

4 CONCLUSION

To be fair, legal controversies in research are surprisingly rare. However, experience tells us that consent can be a "life or death" issue for a biobank. If the proper protocols are not followed, the results of litigation can be devastating. For example, in 2010 the Texas Department of State Health Services was forced under settlement terms to destroy more than five million blood samples collected from newborn babies.[70] Five parents had sued over the failure to obtain consent.[71] Given the potential for this kind of well-publicized outcome, one would expect a more comprehensive response to the biobanking consent dilemma. However, to date, only a few jurisdictions have taken steps to explicitly resolve the legal challenges outlined above.[72]

We believe the biobanking community needs to come to terms with both the reality that the types of consent used in biobanking often do not meet the requirements necessitated by relevant legal norms and, more importantly, the numerous social forces and cultural trends that may be intensifying these unresolved consent issues. That said, we also understand the practical needs that drove the adoption of the modified consent strategies. Researchers, participants, and institutions would all benefit from a defensible, sustainable, and conceptually coherent consent policy. Given the rise in privacy concerns, the increased interest in rights of control, the rapid pace of technological development, and the lack of consensus on preferred consent type, now is the time for policymakers and politicians to clear up the confusion.

---

[69] Natalie T. Boutin et al, "Implementation of electronic consent at a biobank: an opportunity for precision medicine research" (2016) Journal of Personalized Medicine 6(2) 17.

[70] Adam Doerr, "Newborn blood spot litigation: 70 days to destroy 5+ million samples" (*Genomics Law Report*, February 2, 2010) www.genomicslawreport.com/index.php/2010/02/02/newborn-blood-spot-litigation-70-days-to-destroy-5-million-samples/ accessed October 4, 2017.

[71] *Ibid.*

[72] Sirpa Soini, "Finland on a road towards a modern legal biobanking infrastructure" (2013) European Journal of Health Law 20(3) 289.

PART IV

Biobanks, guidelines, and good governance

11. Responsible research and innovation and the advancement of biobanking and biomedical research

Helen Yu[1]

1 INTRODUCTION

One of the core objectives of responsible research and innovation (RRI) is to maximize the value of publicly funded research so it may be returned, to the benefit of society. RRI encourages production of new innovations through societal engagement and collaborative research.[2] It implies close cooperation between all stakeholders involved in the innovation process and requires a setting that supports and fosters collaboration to conduct research with and for society. However, RRI as currently described by the European Commission (the "Commission") can give rise to complex and previously untested issues that challenge the existing legal frameworks on intellectual property and public entitlement to benefits of research.[3] Specifically, the Commission identified open access as one of the key pillars of RRI practice,[4] extending the concept to encompass "open science" and therefore adopting the following definition of open access/open science: "a practice in which the scientific

[1] A version of this chapter was previously published in Journal of Law and the Biosciences (2016), 3(3) 611–35.

[2] European Commission, "Europe 2020: A Strategy for Smart, Sustainable and Inclusive Growth" (Communication) COM (2010) 2020 final; European Commission, "Horizon 2020 EU Framework Programme for Research and Innovation—Responsible Research and Innovation" https://ec.europa.eu/programmes/horizon2020/en/h2020-section/responsible-research-innovation accessed March 5, 2016.

[3] European Commission, "Indicators for Promoting and Monitoring Responsible Research and Innovation" (2015) http://ec.europa.eu/research/swafs/pdf/pub_rri/rri_indicators_final_version.pdf accessed June 4, 2018.

[4] Commission, "on access to and preservation of scientific information" (Recommendation) C(2012) 4890 final.

process is shared completely and in real time."[5] This definition is particularly problematic from a patent law perspective given that, among other factors, a valid patent grant depends on the lack of any prior enabling disclosure in the art.[6] By encouraging the sharing of scientific knowledge "completely and in real time," the Commission's approach to open access/open science, in addition to its endorsement of open innovation, has the potential to essentially foreclose the ability to preserve knowledge for the purposes of obtaining patent protection. Without intellectual property to safeguard investments in R&D, potential stakeholders may be discouraged from contributing to the translation and commercialization of discoveries into innovations if there is no incentive to participate and collaborate. In other words, there is no basis in the law that will incentivize societal engagement and support the *type* of open access/ open science described in RRI policy documents to realize the public benefit objective, unless RRI principles also consider protecting the interests of stakeholders involved in the translation and commercialization of knowledge. The EU has identified research and innovation as key pillars of its strategy to create sustainable growth and prosperity in Europe.[7] If research is to function as an efficient driver of growth, it is imperative that EU policies support the efficient development of academic discoveries into products and technologies that can address societal challenges and achieve socially desirable outcomes. Translation and commercialization are therefore essential mechanisms to ensure research can be transformed into innovations, so that they can be introduced into the market for the benefit of society.

2 RESPONSIBLE RESEARCH AND INNOVATION

The definition of RRI, as adopted by the Commission, is a

> comprehensive approach of proceeding in research and innovation in ways that allow all stakeholders that are involved in the processes of research and innovation at an early stage to obtain relevant knowledge on the consequences of the outcomes of their actions and on the range of options open to them and to effectively evaluate both outcomes and options in terms of societal needs and moral values and to use

---

5 *Supra* note 2, at 32.

6 Article 54(1) of the Convention on the Grant of European Patents, October 5, 1973 1065 UNTS 199.

7 European Commission, "Responsible Research and Innovation—Europe's Ability to Respond to Societal Challenges" (2012) https://ec.europa.eu/research/swafs/pdf/pub_rri/KI0214595ENC.pdf accessed 4 June 2018.

these considerations as functional requirements for design and development of new research, products and services.[8]

The objective is therefore to reduce the risk of societal opposition to new innovations if all stakeholders are involved and consulted throughout the innovation process.[9] The Commission initially identified six key pillars of RRI,[10] and subsequently added two more.[11] Of particular interest and concern to this article is the Commission's approach to the concept of open access/ open science. The growing volume of literature advocating agendas of open access and open science is remarkable and provides a strong argument that commercialization efforts conflict with the free exchange of scientific knowledge and have the potential to jeopardize collaborative research.[12] Numerous organizations have expressly adopted policies that embrace open access and free exchange of scientific information, including, by way of example, the OECD's *Principles and Guidelines for Access to Research Data from Public Funding*,[13] UNESCO's *International Declaration on Human Genetic Data*,[14] the UK's Medical Research Council's Data Sharing Policy,[15] and the Global

---

[8] European Commission, "Options for Strengthening Responsible Research and Innovation—Report of the Expert Group on the State of Art in Europe on Responsible Research and Innovation" (2013) https://ec.europa.eu/research/science-society/document_library/pdf_06/options-for-strengthening_en.pdf accessed June 4, 2018.

[9] Rene Von Schomberg, "A Vision of Responsible Innovation" in Richard Owen, Maggy Heintz, and John Bessant (eds), *Responsible Innovation: Managing the Responsible Emergence of Science and Innovation in Society* (Wiley, 2013).

[10] Commission, *supra* note 6.

[11] *Supra* note 2.

[12] See for example Miriam Bentwich, "Changing the Rules of the Game: Addressing the Conflict between Free Access to Scientific Discovery and Intellectual Property Rights" (2010) 28(2) Nature Biotechnology 137; Matthew Herder, "Choice Patents" (2011) 52(3) IDEA: The Journal of Law and Technology 309; Wei Hong and John P. Walsh, "For Money or Glory? Commercialization, Competition, and Secrecy in the Entrepreneurial University" (2009) 50(1) The Sociological Quarterly 145; Michael A. Heller and Rebecca S. Eisenberg, "Can Patents Deter Innovation? The Anticommons in Biomedical Research" (1998) Science 280, 698–701.

[13] "Full and open access to scientific data should be adopted as the international norm for the exchange of scientific data derived from publicly funded research": www.oecd.org/sti/sci-tech/38500813.pdf accessed May 28, 2018.

[14] Article 18(c) states that "[r]esearchers should endeavour to establish cooperative relationships, based on mutual respect with regard to scientific and ethical matters and … should encourage the free circulation of human genetic data and human proteomic data in order to foster the sharing of scientific knowledge": http://portal.unesco.org/en/ev.php-URL_ID=17720&URL_DO=DO_TOPIC&URL_SECTION=201.html accessed May 28, 2018.

[15] "The MRC expects valuable data arising from MRC-funded research to be made available to the scientific community with as few restrictions as possible so as to

Alliance for Genomics and Health's *Framework for Responsible Sharing of Genomic and Health-Related Data*.[16]

Under an open access policy, researchers are encouraged to freely share knowledge and data quickly to foster scientific progress and meet humanitarian goals.[17] Open access is a means of disseminating research in a timely fashion, with the intention of accelerating scientific discovery and encouraging innovation by reducing barriers and permitting reuse of available materials with few restrictions.[18] The underlying assumption of the concept is that broader participation in the discovery of new knowledge and unrestricted access to knowledge will accelerate the understanding, advancement, and use of science. The Commission decisively extended the concept of open access to embrace open science by encouraging EU member States to make publicly funded research findings and results freely available to the public as a means to support information exchange, collaboration, and communication among stakeholders.[19] At the same time, university imperatives and European Research Area guidelines on intellectual property management in international research collaborations urge researchers to protect the commercial potential of their research by patenting and forming close partnerships with industry to facilitate the translation of knowledge into products.[20] The World Intellectual

---

maximize the value of the data for research and for eventual patient and public benefit. Such data must be shared in a timely and responsible manner": https://mrc.ukri.org/documents/pdf/mrc-data-sharing-policy/ accessed June 7, 2018.

[16] "Seek to make data and research results widely available, including through publication and digital dissemination, whether positive, negative or inconclusive, depending on the nature and use of the data. Dissemination of data and research results should be conducted in a way that both promotes scientific collaboration, reproducibility and broad access to data, and yet minimizes obstacles to data sharing while minimizing harms and maximizing benefits to individuals, families and communities": www.ga4gh.org/ga4ghtoolkit/regulatoryandethics/framework-for-responsible-sharing-genomic-and-health-related-data/ accessed June 7, 2018.

[17] *Supra* note 2.

[18] www.rri-tools.eu/open-access accessed May 28, 2018.

[19] Commission, "on access to and preservation of scientific information" (Recommendation) SWD (2012) 221 final; see also http://ec.europa.eu/programmes/horizon2020/en/h2020-section/open-science-open-access accessed May 28, 2018.

[20] Knowledge Transfer Working Group of the European Research Area Committee, "European Research Area Guidelines on Intellectual Property (IP) Management in International Research Collaboration Agreements between European and Non-European Partners" (2012); see also Timothy Caulfield, "Commercialization Creep" (2012) Policy Options 34, 20–3; Tania M. Bubela and Timothy Caulfield, "Role and Reality: Technology Transfer at Canadian Universities" (2010) 28(9) Trends in Biotechnology 447; C.J. Murdoch and Timothy Caulfield, "Commercialization, Patenting and Genomics: Researcher Perspectives" (2009) 1(2) Genome Medicine 22; Manuel Crespo and Houssine Dridi, "Intensification of University–Industry Relationships

Property Organization also provides guidelines to universities and research organizations on developing policies related to managing intellectual property rights in research findings and academic discoveries.[21] The intention of the guidelines is to facilitate greater collaboration between the research community and industry and "lay the foundation for knowledge-based economic development."[22]

On the one hand, the Europe 2020 strategy calls for "smart growth" by developing the economy on the basis of knowledge and innovation and "inclusive growth" by fostering a high employment economy.[23] This requires an intellectual property policy that incentivizes and supports translation and commercialization in order to develop a knowledge-based economy and facilitate the economic success of industry in order to create jobs and build a sustainable and competitive economy. On the other hand, RRI policy mandates open access and open science to improve knowledge circulation at the expense of intellectual property rights. As a result, an unwanted but likely outcome resulting from this policy tension is "irresponsibility" arising from the difference between the stakeholders in terms of motivation, interests, and interpretation of RRI to achieve their respective objectives.[24] Beyond the affirming sentiments associated with the "concept" of RRI in and of itself, the emerging literature from an academic and policy perspective seems to indicate much confusion and skepticism regarding the practicability of implementing EU-wide RRI practices.[25] The concept of RRI is rapidly evolving and the reality of innovation is that it is a complex collaborative process involving

---

and Its Impact on Academic Research" (2007) 54(1) Higher Education 61; Francis S. Collins, "Reengineering Translational Science: The Time Is Right" (2011) 3(90) Science Translational 1; Elias A. Zerhouni, Charles A. Sanders, and Andrew C. von Eschenbach, "The Biomarkers Consortium: Public and Private Sectors Working in Partnership to Improve the Public Health" (2007) 12(3) The Oncologist 250; Timothy Caulfield, "Sustainability and the Balancing of the Health Care and Innovation Agendas: The Commercialization of Genetic Research" (2003) Saskatchewan Law Review 66, 629.

21 World Intellectual Property Organization, "Guidelines on Developing Intellectual Property Policy for Universities and R&D Organizations" www.wipo.int/about-ip/en/universities_research/ip_policies/.

22 *Ibid.*

23 *Supra* note 1.

24 Jack Stilgoe, Richard Owen, and Phil Macnaghten, "Developing a Framework for Responsible Innovation" (2013) 43 Research Policy 1568, 1580.

25 See for example Vincent Blok and Pieter Lemmens, "The Emerging Concept of Responsible Innovation" Three Reasons Why It Is Questionable and Calls for a Radical Transformation of the Concept of Innovation" in Bert J. Koops and others (eds) *Responsible Innovation 2: Concepts, Approaches, and Applications* (Springer 2015) 19–35; Von Schomberg, *supra* note 8; Lotte Asveld, Jurgen Ganzevles, and Patricia

multiple stakeholders with potentially conflicting interests and agendas, and different ideas on the conceptualization and application of RRI principles to achieve socially responsible objectives. However, despite the lack of consensus on how to operationalize RRI in practice to facilitate innovation, there are at least 12 active international research projects funded in part by the Commission to develop a robust RRI governance framework.[26] Approximately €462 million of public funds will be allocated to research, develop, and implement a policy that the Commission admits is not entirely clear with regard to its feasibility and uptake on an EU level[27]—a concern that is also shared and expressed in the literature.[28]

3 BIOBANKING AND BIOMEDICAL RESEARCH—THE ROLE OF STAKEHOLDERS

Biobanks can be defined as a "collection of biological material and the associated data and information stored in an organized system for a population or a large subset of a population."[29] Biobanks make it possible for researchers to analyze large collections of genetic, genealogical, and health-related data from diverse donors to translate knowledge of the human genome into clinically relevant outcomes for the benefit of public health.[30] Biobanks are therefore an essential resource in a range of clinical and biomedical research purposes, such as epidemiological studies, drug discovery and development, genomics, and personalized medicine.[31] As such, biomedical research derived from biobanks

---

Osseweijer, "Trustworthiness and Responsible Research and Innovation: The Case of the Bio-Economy" (2015) 28(3) Journal of Agricultural and Environmental Ethics 571.

[26] See ProGReSS, 'More RRI Resources' www.progressproject.eu/more-rri-resources/ accessed May 28, 2018.

[27] *Ibid.*

[28] Sarah R. Davies and Maja Horst, "Responsible Innovation in the US, UK and Denmark: Governance Landscapes" in Bert J. Koops and others (eds), *Responsible Innovation 2: Concepts, Approaches, and Applications* (Springer 2015) 37–56.

[29] OECD, "Glossary of Statistical Terms—Biobank" (2007) https://stats.oecd.org/glossary/detail.asp?ID=7220 accessed June 4, 2018.

[30] Madeleine J. Murtagh and others, "Realizing the Promise of Population Biobanks: A New Model for Translation" (2011) 130(3) Human Genetics 333.

[31] Janet E. Olson and others, "Biobanks and Personalized Medicine" (2014) 86(1) Clinical Genetics 50; Gert-Jan B. van Ommen and others, "BBMRI-ERIC as a Resource for Pharmaceutical and Life Science Industries: The Development of Biobank-Based Expert Centres" (2015) 23(7) European Journal of Human Genetics 893; Thane Kreiner and Stefan Irion, "Whole-Genome Analysis, Stem Cell Research, and the Future of Biobanks" (2013) 12(5) Cell Stem Cell 513; Naomi Allen and others, "UK Biobank: Current Status and What It Means for Epidemiology" (2012) 1(3) Health Policy and Technology 123.

has immense potential for the production of innovations that can be used to advance healthcare. Whether products or processes identified in the context of biobanking should be or are legally eligible for patent protection is the subject of much debate.[32] While human biological materials (HBM) are considered natural products isolated from the human body and are therefore not patentable as discoveries of natural phenomena, according to the European Commission directive on the legal protection of biotechnological inventions, "biological material isolated from its natural environment or produced by means of a technical process may be the subject of an invention even if it previously occurred in nature."[33] As such, the application of research and data derived from HBM could result in patentable inventions. Some scholars argue that intellectual property rights serve as a tool to protect the substantial investments made in research projects using HBM and related biobank data,[34] whereas others argue that intellectual property rights should not be granted on research results derived from HBM.[35] Arguably, the "bigger picture" concern is that of donors regarding the commercialization of innovations arising from research derived from HBM, given that trust has been identified as a key predictor of attitude and intention of participants to donate and participate in biobank research.[36] The ethicolegal tension between the voluntary and often altruistic intentions of donors and the profit motives associated with commercialization attracts much concern and criticism.[37] Biobanks can provide crucial platforms for commer-

---

32 See for example Kathinka Evers, Joanna Forsberg, and Mats Hansson, "Commercialization of Biobanks" (2012) Biopreservation and Biobanking, 45–7; Saminda Pathmasiri, Mylene Deschênes, Yann Joly, Tara Mrejen, Francis Hemmings, and Bartha Maria Knoppers, "Intellectual Property Rights in Publicly Funded Biobanks: Much Ado About Nothing?" (2011) Nature Biotechnology 29(4), 319–24; Timothy Caulfield, "Reflections on the Gene Patent War" (2011) Clinical Chemistry 57(7), 977–9.

33 See Council Directive 1998/44/EC of 6 July 1998 on the legal protection of biotechnological inventions [1998] OJ L213/13, arts 3 and 5.

34 See for example Julien Pénin and Jean-Pierre Wack, "Research Tool Patents and Free-Libre Biotechnology: A Suggested Unified Framework" (2008) 37(10) Research Policy 1909.

35 See for example Christopher Heaney and others, "The Perils of Taking Property Too Far" (2009) 1 Stanford Journal of Law, Science & Policy 46; Caulfield, *supra* note 31.

36 Christine Critchley, Dianne Nicol, and Margaret Otlowski, "The Impact of Commercialisation and Genetic Data Sharing Arrangements on Public Trust and the Intention to Participate in Biobank Research" (2015) 18(3) Public Health Genomics 160. See also Recital 26 of Directive 1998/44/EC, *supra* note 32, which requires patent applications for inventions based on HBM to provide donors of such HBM the opportunity to "[express] free and informed consent".

37 See for example Daryl Pullman and others, "Personal Privacy, Public Benefits and Biobanks: A Conjoint Analysis of Policy Priorities and Public Perceptions" (2012)

cially valuable research but the highly personal nature of HBM, and that fact that often personal data must accompany the sample to be useful, intensifies the need to adhere to RRI principles with respect to the research, development, and commercialization of innovations derived from biobank data and donor samples. In keeping with RRI principles and objectives to maximize the value of publicly funded research, the translation and commercialization of innovations arising out of research derived from biobanks should endeavor to balance ethical concerns with the best possible use of HBM for the benefit of the public.[38] The need for a balance of interests is also articulated by the Human Genome Organization, which states: "Knowledge useful to human health belongs to humanity. Human genomic databases are a public resource. All humans should share in and have access to the benefits of databases."[39] At the same time, "[r]esearchers, institutions, and commercial entities have a right to a fair return for intellectual and financial contributions to database [but] any fees should not restrict the free flow of scientific information and equitable access."[40] However, as often stated in the literature, "what is in the best interests of the public in the context of [publicly] funded biobanks is far from obvious."[41]

Commercialization may also result in tension between the interests of industry and those of researchers involved in biobanking. Collaborations between industry and academia incentivized by commercial interests may have the effect of compromising research integrity.[42] For example, studies have shown

---

14(2) Genetics in Medicine 229; Herbert Gottweis, George Gaskell, and Johannes Starkbaum, "Connecting the Public with Biobank Research: Reciprocity Matters: (2011) 12(11) Nature Reviews Genetics 738; Tore Nilstun and Goran Hermeren, "Human Tissue Samples and Ethics—Attitudes of the General Public in Sweden to Biobank Research" (2006) 9(1) Medicine, Health Care and Philosophy 81; Timothy Caulfield, Christen Rachul, and Erin Nelson, "Biobanking, Consent, and Control: A Survey of Albertans on Key Research Ethics Issues" (2012) 10(5) Biopreservation and Biobanking 433; Zubin Master and others, "Cancer Patient Perceptions on the Ethical and Legal Issues related to Biobanking" (2013) 6(1) BMC Medical Genomics 8; A.A. Lemke and others, "Public and Biobank Participant Attitudes toward Genetic Research Participation and Data Sharing" (2010) 13(6) Public Health Genomics 368; Susan B. Trinidad and others, "Genomic Research and Wide Data Sharing: Views of Prospective Participants" (2010) 12(8) Genetics in Medicine 486.

[38] Evers, Forsberg, and Hansson *supra* note 31.

[39] HUGO Ethics Committee, "Statement on Human Genomic Database" (2002) Recommendation 1, www.hugo-international.org/Resources/Documents/CELS_Statement-HumanGenomicDatabase_2002.pdf accessed May 28, 2018.

[40] *Ibid*, Recommendation 6.

[41] Pathmasiri and others, *supra* note 31.

[42] Justin E. Bekelman, Yan Li, and Cary P. Gross, "Scope and Impact of Financial Conflicts of Interest in Biomedical Research: A Systematic Review" (2003)

that "industry-sponsored research" in the field of biomedical research tends to support and promote proindustry conclusions,[43] which may further undermine public perceptions of stakeholder involvement in biomedical research. If participants view stakeholder involvement as introducing a profit motive into what is otherwise an act of public good, there is a risk that the public may refuse to donate HBM or elect to withdraw their samples.[44] In the eyes of potential donors, public funding of biobanks connotes a common good, scientific and public health benefits, and values of sharing and trust, whereas notions of profit, private interest, and economic benefit conjure mistrust around privately funded biobanks.[45] Financial incentives may work contrary to the ambition of using biobank samples and associated data to address public health problems by impeding other research collaborations and by modifying the research agenda to satisfy the commercial interests of industry at the expense of addressing important health problems.[46] The literature is equivocal as to whether the public's objection is principled and directed at commercialization in the biomedical field as such, or if the opposition is pragmatic and directed at the possible unjust or exploitative consequences of commercialization strategies. In any event, challenges related to biobanking and biomedical research have emerged whereby ethical, legal, social, cultural, and economic considerations must be taken into account when formulating policy regarding the advancement of biobanking and biomedical research.

Stakeholders involved in the research, development, translation, and commercialization of innovations derived from biobank research collaborate with each other for their own reasons: For example: (i) donors want to protect their privacy and benefit from the research that uses their samples; (ii) researchers want free and open access to knowledge and data to foster scientific progress;

289(4) Journal of the American Medical Association 454; Joel Lexchin and others, "Pharmaceutical Industry Sponsorship and Research Outcome and Quality: Systematic Review" (2003) 326 BMJ 1167; Mohit Bhandari and others, "Association Between Industry Funding and Statistically Significant Pro-Industry Findings in Medical and Surgical Randomized Trials" (2004) 170(4) Canadian Medical Association Journal 477.

[43] Bekelman and others, *ibid*.

[44] Dianne Nicol and Christine Critchley, "Benefit Sharing and Biobanking in Australia" (2012) 21(5) Public Understanding of Science 534; Caulfield, Rachul, and Nelson, *supra* note 36.

[45] Maurizio Onisto, Viviana Ananian, and Luciana Caenazzo, "Biobanks between Common Good and Private Interest: The Example of Umbilical Cord Blood Private Biobanks" (2011) 5(3) Recent Patents on DNA & Gene Sequences 166.

[46] Klaus Hoeyer, "Trading in Cold Blood?" in Peter Dabrock, Jochen Taupitz, Jens Ried (eds), *Trust in Biobanking: Dealing with Ethical, Legal and Social Issues in an Emerging Field of Biotechnology* (Springer 2012) 21–41.

(iii) industry wants to invest in research that will lead to commercial benefits; (iv) universities seek to discover and disseminate knowledge as well as to attract public and private funding for further research; (v) government wants to support research that will drive socioeconomic growth and create greatest impact; and (vi) the public is interested in a return of benefit from the tax dollars invested in basic research. See Figure 11.1.

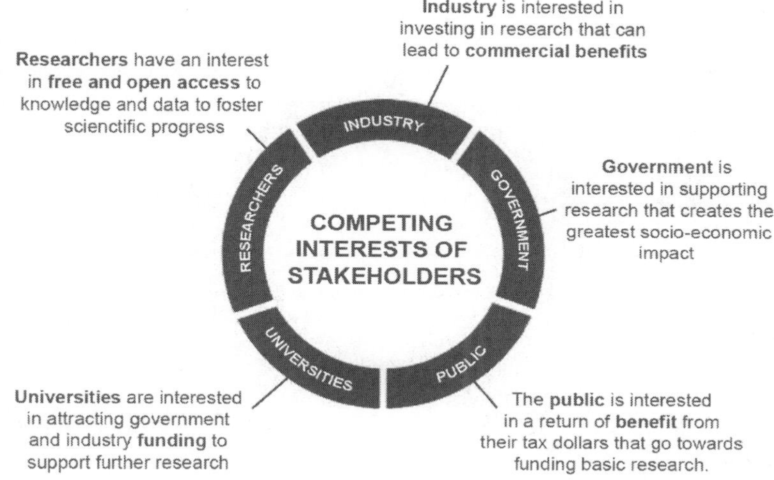

Figure 11.1 Stakeholder interests

Although the advancement of biomedical research derived from biobank data involves the participation of a number of stakeholders, each with different interests and motivations, the parties share at least one common desired outcome: the discovery of new innovations for the advancement of healthcare. To realize the public benefit objective of biobank-related research, the interests of all stakeholders involved in the translation and commercialization of knowledge must be considered.

3.1 Donors and the Public

There is much debate in the literature about the ethicolegal tension between the economics of translating and commercializing innovations derived from biobank resources and the scientific value of HBM for the advancement of biomedical research.[47] Public trust issues tend to be raised in association with the commercialization of innovations derived from biobanks, which must be properly addressed to inspire public support for the advancement of human health.[48] Industry involvement can raise questions about financial motives, thus compromising the integrity of the biobank and associated research in the eyes of the public. Some argue that commercializing biobank resources threatens to undermine the altruistic donation of individual donors,[49] and that commercial interests may direct research toward the needs of industry as opposed to the scientific and public good.[50] There is evidence suggesting that donor trust and support in biobanking research significantly decreases if researchers are involved with industry or government as opposed to universities.[51] Numerous guidelines and recommendations specifically recognize the ethical responsibility of researchers to inform donors about potential commercial applications resulting from research on donor HBM.[52] Donor participation and right to

---

[47] See for example Danijela Budimir and others, "Ethical Aspects of Human Biobanks: A Systematic Review" (2011) 52(3) Croatian Medical Journal 262; Klaus Hoeyer, "The Ethics of Research Biobanking: A Critical Review of the Literature" (2008) 25(1) Biotechnology and Genetic Engineering Reveiws 429; Richard Tutton, "Biobanking: Social, Political and Ethical Aspects" (2010) Encyclopedia of Life 1.

[48] David B. Resnik, "Scientific Research and the Public Trust" (2011) 17(3) Science and Engineering Ethics 399; Zubin Master and David B. Resnik, "Hype and Public Trust in Science" (2013) 19(2) Science and Engineering Ethics 321, which states that "'[p]ublic trust' is not a static or easily quantifiable concept. Rather, it is relational, ongoing, and changing … [T]he 'public' is not a homogenous entity that speaks with one voice: there are many different groups that comprise 'the public', and these groups may differ in their trust of scientists. These relationships of trust may be affected by a number of different factors and change at different periods of time."

[49] Catherine Waldby, "Biobanking in Singapore: Post-developmental State, Experimental Population" (2009) 28(3) New Genetics and Society 253; Daryl Pullman and others, "Personal Privacy, Public Benefits, and Biobanks: A Conjoint Analysis of Policy Priorities and Public Perceptions" (2012) 14(2) Genetics In Medicine 229; Herbert Gottweis and others, *supra* note 36, at 738–9; Dianne Nicol and Christine Critchley, "Benefit Sharing and Biobanking in Australia" (2012) 21 Public Understanding of Science 534, 555; Caulfield, Rachul, and Nelson, *supra* note 36.

[50] Robert Mitchell and Catherine Waldby, "National Biobanks: Clinical Labor, Risk Production, and the Creation of Biovalue" (2010) 35(3) Science, Technology & Human Values 330.

[51] Caulfield, Rachul, and Nelson, *supra* note 36; Zubin and others, *supra* note 36; Michael Clemence and others, *Wellcome Trust Monitor Wave 2: Tracking Public Views on Science, Biomedical Research and Science Education* (London 2013); see also The Swinburne National Technology and Society Monitor (proposed July 30, 2012) www.swinburne.edu.au/lss/spru/spru-monitor.html accessed May 28, 2018.

[52] Organisation for Economic Co-Operation and Development, OECD Guidelines on Human Biobanks and Genetic Research Databases (OECD 2009); World Health Organization, Guideline for Obtaining Informed Consent for the Procurement and

withdraw might be at stake if the public does not associate commercialization with efforts in the interest of the public good.

Furthermore, underlying this commercialization tension is the financial commitment required to ensure the long-term sustainability of biobanks, in order to support ongoing biomedical research for the benefit of public health. Evidence indicates that publicly funded biobanks are concerned with ensuring long-term funding in response to financial pressures on public funding,[53] making partnerships with stakeholders a pragmatic means to secure financial security. In order to maintain quality and scientific efficacy, some biobanks have resorted to operating like a business enterprise in order to support continued scientific endeavors.[54] In order to benefit from biomedical research, the sustainability of biobanks as a resource on which such research is based needs to be secured. However, introducing private funding and partnerships to existing publicly funded biobanks can give rise to various policy and legal concerns. Public trust declines,[55] and there is fear that the involvement of stakeholders may limit or prevent the sharing and return of biobank resources

---

Use of Human Tissues, Cells, and Fluids in Research (WHO 2003); Council for International Organizations of Medical Sciences, International Ethical Guidelines for Biomedical Research Involving Human Subjects (CIOMS 2002), Council of Europe, Treaty Series—No 195 Additional Protocol to the Convention on Human Rights and Biomedicine, concerning Biomedical Research (COE 2005).

[53] Gail E. Henderson and others, "Characterizing Biobank Organizations in the US: Results From a National Survey" (2013) 5(1) Genome Medicine 3; R. Jean Cadigan and others, "Neglected Ethical Issues in Biobank Management: Results from a US Study" (2013) 9(1) Life Sciences, Society & Policy 1; Aaro Tupasela and Neil Stephens, "The Boom and Bust Cycle of Biobanking—Thinking through the Life Cycle of Biobanks" (2013) 54 Croatian Medical Journal 501; Saminda Pathmasiri and others, "Intellectual Property Rights in Publicly Funded Biobanks: Much Ado about Nothing?" (2011) 29(4) Nature Biotechnology 319; Ingeborg Meijer, Jordi Molas-Gallart, and Pauline Mattsson, "Networked Research Infrastructures and their Governance: The Case of Biobanking" (2012) 39(4) Science & Public Policy 491.

[54] Sandra A. McDonald and others, "Fee-for-Service as a Business Model of Growing Importance: The Academic Biobank Experience" (2012) 10(5) Biopreservation and Biobanking 421.

[55] 32 percent of respondents trusted scientists working with industry, 34 percent trusted scientists working with government, and 66 percent trusted university scientists. See Clemence and others, *supra* note 50; Caulfield and others, *supra* note 36; Christine Critchley and Lyn Turney, "Understanding Australians' Perceptions of Controversial Scientific Research" (2004) 2(2) Australian Journal of Emerging Technologies & Society 82; Christine R. Critchley and Dianne Nicol, "Understanding the Impact of Commercialization on Public Support for Scientific Research: Is It About the Funding Source or the Organization Conducting the Research" (2011) Public Understanding of Science 347.

and the results derived therefrom.[56] Public expectations that society is entitled to results derived from publicly funded research, regardless of stakeholder participation, can lead to the creation of biobanking policies that may be ethically or legally contentious.[57]

The negative attitude toward commercial entities' involvement in biobanks and biobanking research is in part associated with questions regarding the degree to which research is being done ethically and primarily for the public good, as opposed to in support of commercial interests.[58] However, this attitude may be based in part on misconceptions. There is evidence indicating that the public generally has limited understanding of the translational process and the contributions required from stakeholders to translate and commercialize basic research into innovations.[59] In a recent UK report on public attitudes to commercial access to health data, it was concluded that the public knows very little about some key areas, including how the commercial sector contributes to healthcare, the role of universities and industry in the development of health and medical research, and the processes through which medical and scientific research is carried out to produce drugs and therapies.[60] The public also wanted safeguards put in place to regulate profit motives by creating independent scrutiny and control.[61] To benefit from the publicly funded research derived from donated HBM, industry needs to commercialize and translate early stage discoveries into socially beneficial innovations, *but* industry will only participate if there is a financial incentive to do so. To promote effective uptake of publicly funded research, the public needs to be informed of the realities of the translation process. Members of the public need to critically assess their own preconceptions and know that their participation in collaboration with other

---

56 Christine R. Critchley, "Public Opinion and Trust in Scientists: The Role of the Research Context and the Perceived Motivation of Stem Cell Researchers" (2008) 17(3) Public Understanding of Science 309.

57 Timothy Caulfield, "Biobanks and Blanket Consent: The Proper Place of the Public Good and Public Perception Rationales" (2007) 18(2) King's Law Journal 209.

58 Nilstun and others, *supra* note 36; Lemke and others, *supra* note 36; Susan Brown Trinidad and others, *supra* note 36.

59 Andy Haines, Shyama Kuruvilla, and Matthias Borchert, "Bridging the Implementation Gap between Knowledge and Action for Health" (2004) 82(10) Bulletin of the World Health Organization 724; Evers, Forsberg, and Hansson, *supra* note 31; Jennifer L. Baumbusch and others, "Pursuing Common Agendas: A Collaborative Model for Knowledge Translation between Research and Practice in Clinical Settings" (2008) 31(2) Research In Nursing And Health 130.

60 Ipsos MORI Social Research Institute, "The One-Way Mirror: Public Attitudes to Commercial Access to Health Data," prepared for the Wellcome Trust (2016) www.ipsos.com/sites/default/files/publication/5200-03/sri-wellcome-trust-commercial-access-to-health-data.pdf accessed June 4, 2018.

61 *Ibid.*

stakeholders, including industry, is required in order to facilitate the creation and introduction of socially beneficial innovations into the market. A number of research groups have attempted to develop best practices in biobank governance and engage the public in a wider debate about and understanding of biobanking and biomedical research.[62] There is evidence indicating that the more informed people are, the more likely they are to approve of use of their HBM for the advancement of public health, including the involvement of industry.[63]

3.2 Industry

Due to the increasing cost of R&D and budgetary and funding challenges in the public research sector, collaborative partnerships between industry and public research organizations, such as universities, may be a pragmatic means to pool resources, share information, and reduce duplication efforts in order to optimize the impact of research and derisk the development of basic research.[64] If collaborative partnerships are to be supported as a means to facilitate the translation process, public research organizations and industry must find ways to forge closer ties. Alliances naturally involve risk and concerns, such as how to preserve the core values of academia while providing industry with the incentive required to justify investment in basic research.[65] Industry, and particularly the pharmaceutical industry, has long maintained that patents are crucial to the financial viability of continued R&D.[66] Given the costly and

---

[62] See for example Karolinska Institute and BBMRI, Education and Lectures on Biobanking http://ki.se/en/research/education-and-lectures-on-biobanking accessed June 4, 2018; UKCRC Tissue Directory and Coordination Centre stakeholder engagement www.biobankinguk.org accessed May 28, 2018.

[63] *Supra* note 59, at 107.

[64] See for example Nuala Moran, "Public Sector Seeks to Bridge 'Valley of Death'" (2007) 25 Nature Biotechnology 266; Paul K. Owens and others, "A Decade of Innovation in Pharmaceutical R&D: The Chorus Model" (2015) 14(1) Nature Reviews Drug Discovery 17.

[65] John P. Walsh, Wesley M. Cohen, and Charlene Cho, "Where Excludability Matters: Material versus Intellectual Property in Academic Biomedical Research" (2007) 36(8) Research Policy 1184.

[66] Roy Levy, *FTC Bureau of Economics Staff Report—The Pharmaceutical Industry: A Discussion of Competitive and Antitrust Issues in an Environment of Change* (DIANE Publishing 1999), which concluded that an estimated 65 percent of the drugs on the market would not have been developed at all absent patent protection; see also Edwin Mansfield, "Patents and Innovation: An Empirical Study" (1986) 32(2) Management Science 173, which found that up to 90 percent of pharmaceutical inventions would not have been developed without patents.

risky nature of translating discoveries into innovations,[67] it is understandable that industry advocates for intellectual property protection as an incentive to derisk the development of early stage research and to be rewarded for their investment should efforts lead to commercialization. Aside from financial incentives, industry also views patent rights as a means to facilitate collaboration and induce R&D investment by allowing the parties with the best knowledge of a particular market or technology to make decisions on how to manage the risks and protect their innovation.[68] In other words, the transactional function of patents (as opposed to the proprietary rights) can itself be seen as an incentive to facilitate collaboration by providing a system by which the parties can organize collaborative R&D in the most efficient manner based on their unique knowledge of market dynamics.[69] By reducing transactional hazards associated with creating the relationship between stakeholders, discoveries have a greater likelihood of being translated into innovations for the benefit of the public. As long as industry involvement in the translation of basic research derived from HBM is necessary to achieve this outcome, financial motives associated with commercialization are an undeniable fact, whether or not there is empirical evidence to support the notion that patents are essential to stimulate innovation.

---

[67] However, there is much debate and conflicting reports over the actual cost of drug development. See for example Donald W. Light and Rebecca Warburton, "Demythologizing the High Costs of Pharmaceutical Research" (2011) 6(1) Journal Biosocieties 34.

[68] William M. Landes and Richard A. Posner, *The Economic Structure of Intellectual Property Law* (Harvard University Press 2009); James Langenfeld, "Intellectual Property and Antitrust: Steps toward Striking a Balance" (2001) Case Western Reserve Law Review 52, 91. Because parties never have perfect information when entering into a collaboration, uncertainties and information asymmetries manifest themselves as transaction costs when parties use best efforts to negotiate agreements governing their relationship. See Paul L. Joskow, "Transaction Cost Economics, Antitrust Rules and Remedies" (2002) 18(1) Journal of Law, Economics and Organization 95 and Herbert Hovenkamp, "Harvard, Chicago, and Transaction Cost Economics in Antitrust Analysis" (2012) 57(3) The Anitrust Bulletin 613.

[69] Nancy T. Gallini and Suzanne Scotchmer, "Intellectual Property: When Is It the Best Incentive System?" in Adam B. Jaffe, Josh Lerner, and Scott Stern (eds), *Innovation Policy and The Economy* (vol. 2, MIT Press 2002) 51–78; Bronwyn H. Hall and Dietmar Harhoff, *Recent Research on the Economics of Patents* (National Bureau of Economic Research working paper 2012).

3.3 Public Research Organization and Researchers

Over the past few decades, the position of universities and public research organizations in the market has evolved considerably.[70] Gradually, they have become more actively involved in the process of transferring technology to industry, since placing publicly funded research in the public domain is no longer seen as sufficient to generate the full benefits of innovation.[71] The overall evolution of the economy toward a knowledge economy creates an important incentive for constant innovation and exploitation of new technologies, which very naturally brings together academia on the one hand and industry on the other.[72] Policy changes to encourage collaboration between academia and industry to commercialize knowhow allow academia to own and license patents for inventions derived from publicly funded research.[73] Industry benefits from having access to cutting edge research and discoveries and academia benefits from receiving royalties to fund further R&D, to compensate for budgetary cuts to public funding.[74] Research also indicates that university–industry collaborations positively affect academic research performance in terms of patenting and publication activities.[75]

However, despite a shared commitment (and government support to foster university–industry collaborations), significant cultural obstacles stand in the way of successful partnerships between researchers and industry.[76] Academics speak

---

[70] Bart Van Looy and others, "Entrepreneurial Effectiveness of European Universities: An Empirical Assessment of Antecedents and Trade-Off" (2011) 40(4) Research 553, which stated that since the late 1970s, many countries have changed their legislation and created support mechanisms to encourage interaction between universities and firms, including through technology transfer.

[71] "Introduction" in Mike Wright and others (eds), *Academic Entrepreneurship in Europe* (Edward Elgar 2007).

[72] Petra Andries and Koenraad Debackere, "Adaptation and Performance in New Businesses" (2007) 29(1–2) Small Business Economics 81.

[73] Henry W. Chesbrough, "Business Models and Managing Intellectual Property" in *Open Innovation* (Harvard Business School Press 2003); David Roessner and others, "The Economic Impact of Licensed Commercialized Inventions Originating in University Research" (2013) 42(1) Research Policy 23.

[74] Joseph Friedman and Jonathan Silberman, "University Technology Transfer: Do Incentives, Management, and Location Matter?" (2003) 28(1) Journal of Technology Transfer 17.

[75] Bart Van Looy and others, "Combining Entrepreneurial and Scientific Performance in Academia: Towards a Compounded and Bi-Directional Matthew-Effect" (2004) 33(3) Research Policy 425; Bart Van Looy, Julie Callaert, and Koenraad Debackere, "Publication and Patent Behavior of Academic Researchers: Conflicting, Reinforcing or Merely Co-Existing?" (2006) 35(4) Research Policy 596.

[76] Kenneth I. Kaitin, "Translational Research and the Evolving Landscape for Biomedical Innovation" (2012) 60(7) Journal of Investigative Medicine 995.

the language of science and industry speaks the language of business. Academics are generally motivated by research and publication,[77] and industry is motivated by commercial interests. In other words, conflicting objectives create reluctance among the parties to align too closely. Academia is a rich source of basic research and discovery but lacks the funding and translational expertise regarding how new therapies reach the market. Industry specializes in translational activities and procedures required to convert early stage research into new therapies but lacks the competency and resources to conduct basic research. It is the mutual desire to develop and deliver new treatments, therapies, and medicines that drives the collaboration and development of partnerships to bridge the translational gap. Furthermore, the relationship is mutually beneficial: academia requires funding to conduct research; industry can provide funding. Industry requires innovative research to commercialize; academia can provide cutting-edge early stage research and discoveries.

3.4 Government

The European Strategy Forum on Research Infrastructure has identified biobanks as one of the main priority research areas.[78] The creation of the pan-European Biobanking and Biomolecular Resources Research Infrastructure (BBMRI) is intended to enable the identification of new targets for therapy and reduce attrition in drug discovery and development by facilitating the translation of basic research discoveries into the development of innovative strategies for the prevention, diagnosis, and treatment of diseases of particular relevance to the EU and alleviating the associated medical and economic burden.[79] The Commission also recognizes that sound governance of biobanks is one of the most important challenges to ensure that biobanking is conducted ethically and responsibly.[80]

---

[77] There is literature on a phenomenon known as "academic entrepreneurs" where scientists are interested in creating a spinoff company around their research to develop the commercial potential and utility of publicly funded research and "entrepreneurial academics" where scientists pursue research interests in an entrepreneurial setting. See for example Martin Meyer, "Academic Entrepreneurs or Entrepreneurial Academics? Research-based Ventures and Public Support Mechanisms" (2003) 33(2) R&D Management 107; John Egan, Ceri Williams, and Josephine Dixon-Hardy, *When Science Meets Innovation: A New Model of Research Translation* (International Society for Professional Innovation Management 2013).

[78] Commission, *A Vision for Strengthening World Class Research Infrastructures on the ERA: Report of the Expert Group on Research Infrastructures* (Brussels 2010).

[79] *Ibid.*

[80] European Commission, "Biobanks for Europe—A Challenge for Governance" (2012) www.coe.int/t/dg3/healthbioethic/Activities/10_Biobanks/biobanks_for_Europe.pdf accessed June 4, 2018.

In addition to its interest in addressing societal concerns and challenges, the Commission has an interest in maintaining the economic health and prosperity of the EU. According to the Commission, research and development could create 3.7 million jobs and increase annual GDP by close to €800 billion by 2025.[81] At the heart of the Europe 2020 strategy is the adoption of a more strategic approach to innovation, without reference to any particular fields of research, that will contribute to Europe's competitiveness and provide an associated increase in job numbers.[82] From the Commission's perspective, the expected positive impact of research and innovation depends largely on the ability to fund research to regain Europe's economic foundation and achieve prosperity.[83] The underlying assumption is that research and innovation will lead to socioeconomic growth (that is, more and faster innovation means prosperity, job creation, and the overall betterment of society). The economic prosperity and sustainable growth that innovation is expected to yield is therefore dependent upon stakeholders' ability to collaborate effectively on the translation and commercialization of publicly funded research. The socioeconomic growth sought by government will only be realized if industry can successfully commercialize innovations, which will lead to company creation, job creation, increased tax revenues, and societal uptake of innovations to address societal challenges. Successful commercialization of innovations derived from publicly funded research also demonstrates why government should continue to fund basic research and motivates it to do so. Government will provide public funding to universities and research institutions if there is evidence (that is, successful commercialization) that the funds will be applied toward research with the potential to address societal challenges and drive economic growth.[84]

4 OPEN INNOVATION AND INTELLECTUAL PROPERTY RIGHTS

The literature largely supports the contention that competition on the market is fostered by openness and access to knowledge.[85] It is widely recognized that

---

[81] Commission, "Europe 2020 Flagship Initiative Innovation Union" (Communication) SEC (2010) 1161.

[82] *Supra* note 1.

[83] René Von Schomberg, "Prospects for Technology assessment in a Framework of Responsible Research and Innovation" in Marc Dusseldorp and Richard Beecroft (eds), *Technikfolgen abschätzen lehren: Bildungspotenziale transdisziplinärer Methoden* (Springer 2012) 39–61.

[84] Timothy Caulfield, Shawn H.E. Harmon, and Yann Joly, "Open Science Versus Commercialization: A Modern Research Conflict?" (2012) 4(2) Genome Medicine 17.

[85] See for example Carl Shapiro, "Competition and Innovation: Did Arrow Hit the Bull's Eye?" in Josh Lerner and Scott Stern (eds), *The Rate and Direction of Inventive*

making research results more accessible contributes to better and more efficient science and innovation in the public and private sectors.[86] Although there is a clear link between open science and open innovation, contrary to popular understanding, they are not the same thing.[87] Science has the purpose of developing knowledge by adding theoretical or empirical insights. "Open science" essentially advocates for freely sharing scientific knowledge at the earliest practical point in the discovery process to accelerate the advancement of science.[88] Innovation, meanwhile, has the purpose of transforming knowledge for the purpose of bringing new products or technologies to market. "Open innovation" therefore advocates for the purposive use of available knowledge to accelerate the introduction of new innovations into the market. Both open science and open innovation encourage the acceleration of knowledge through a process of sharing, but that is where the similarities end. Because scientific findings have the potential to be translated into innovations, it is necessary to understand the link between open science and open innovation to determine whether open science will necessarily lead to open innovation.

Despite advances in biomedical research in the recent past and the push for open science/open access, the potential of early stage discoveries and their best application is often unknown or unclear at the outset.[89] Significant time can elapse between discovering knowledge and putting that knowledge into practical use for the benefit of society.[90] For example, it took more than a decade from the discovery of light-activated compounds in the sap of cow parsley to translate research in photodynamic therapy into the drug VISUDYNE for the treatment of blood vessel disorders in the eye.[91] The journey from foundational research of photosensitizer chemicals at a department of botany to discovery of the potential of photo-activated chemicals for therapeutic purposes in a biomedical setting was not immediately apparent to the researchers that eventually enabled the development of VISUDYNE. Translating scientific knowledge into useful applications involves incentives and mechanisms dif-

---

Activity Revisited (University of Chicago Press 2011) 361–404; Henry Chesbrough, "From Open Science to Open Innovation" (2015) www.fosteropenscience.eu/sites/default/files/pdf/1798.pdf accessed June 4, 2018.

86 See for example Sascha Friesike and others, "Opening Science: Towards an Agenda of Open Science in Academia and Industry" (2015) 40(4) Journal of Technology Transfer 581.

87 Roessner, *supra* note 72, at 38.

88 Friedman and Silberman, *supra* note 73, at 18.

89 Chesbrough, *supra* note 84, at 7.

90 Asher Mullard, "New Drugs Cost US$2.6 Billion to Develop" (2014) 877 Nature Reviews Drug Discovery; 47, at 40; 48, at 3.

91 *Mass. Eye & Ear Infirmary v QLT Phototherapeutics, Inc.* (2006) 412 F.3d 215, 221–2 (1st Cir. 2005), cert. denied, 126 S. Ct. 2292.

ferent from those at the discovery and research phase. Academic researchers and scientists are motivated by the pursuit of knowledge and recognition as a result of discoveries during the scientific process.[92] Open access/open science serves to facilitate the pursuit and dissemination of knowledge for the advancement of science. Translating scientific knowledge into useful applications to the benefit of the public usually involves the participation of others in the innovation process, such as industry and translational scientists involved in the application and integration of available knowledge into new innovations, which typically introduces a financial motive into the innovation process.[93] The translation of early stage biomedical research where the clinical and commercial potential of the discovery is unknown involves substantial risks and large investments.[94] Before investing in the development of early stage discoveries, industry needs to evaluate whether there are any third party rights preventing the development and subsequent commercialization of the new innovation. If the discovery is already protected by existing intellectual property rights, an assessment of the scope of those rights must be made to determine if the patent claims adequately protect the new innovation, and if not, whether there are any disclosure problems preventing subsequent patent filings. An infringement and validity assessment must also be conducted. Common business sense dictates that any investor will demand that there be some protection for its investment and an assurance that there are no legal consequences preventing the development and commercialization of the innovation. In other words, open science at the discovery phase does not necessarily lead to translation of knowledge into innovations. Intellectual property rights are therefore critical in incentivizing and inducing industry to undertake the risk of development and invest in the commercialization of knowledge to introduce new innovations from the laboratory to the market.

---

92 Paula E. Stephan, "The Economics of Science" (1996) 34(3) Journal of Economic Literature 1199; Katherine W. McCain, "Communication, Competition, and Secrecy: The Production and Dissemination of Research-Related Information in Genetics" (1991) 16(4) Science, Technology and Human Values 491.

93 Helen W.H. Yu, "Bridging the Translational Gap: Collaborative Drug Development and Dispelling the Stigma of Commercialization" (2016) 21(2) Drug Discvoery Today 299.

94 It has been reported that on average, only 3 in 10 new pharmaceutical products generate revenues equal to or greater than average R&D costs. See Henry G. Grabowski, John M. Vernon, and Joseph A. DiMasi, "Returns on Research and Development for 1990s New Drug Introductions" (2002) 20(3) Pharmacoeconomics 11, and Joseph A. DiMasi and Henry G. Grabowski, "R&D Costs and Returns to New Drug Development: A Review of the Evidence" in *The Oxford Handbook of the Economics of the Biopharmaceutical Industry* (Oxford University Press 2012) 21–46.

Because of the inventive step and novelty requirement for patentability, open access and open science during the scientific discovery process without the assertion of intellectual property may preclude subsequent patent protection of new innovations. The rise of open science and the proliferation of online resources, combined with the accessibility offered by the internet, means both the public and patent offices now have easy access to a wealth of knowledge. Freely, completely, and immediately sharing discoveries and research findings arising from the scientific process through typical dissemination channels such as publications and conferences may inadvertently destroy the novelty and/or inventiveness of future innovations.[95] Novelty and inventiveness searches from patent offices include journals and trade literature as well as patents and patent applications. With the bias of perfect hindsight, open science may contribute an abundance of thorough knowledge to the state of the art that independently or cumulatively prevents the grant of a patent. Theoretically, knowledge arising from the development and translation phase of the innovation process relating to the practical application of scientific discoveries should be protected by intellectual property, while background knowledge that supports the application remains open to the scientific community. However, the line between basic research and applied research is not always crystal clear from a patentability perspective (and with perfect hindsight). Referring back to the VISUDYNE example, does research into the potential of photo-activated chemicals for therapeutic purposes constitute background knowledge, therefore rendering the discovery that the light-activated compound could be used to activate drugs accumulated in a particular area obvious? Widely distributed knowledge through open science/open access means there is a serious risk that the prior art will prevent patents from being granted on the application of scientific discoveries. The delineation between early stage research (where open science is encouraged to foster the advancement of science) and applied research (where the preservation of intellectual property rights becomes relevant to preserve translation and commercialization potential) is particularly difficult for scientists to determine—not that they are trained to do so; nor should they be burdened with such a role.

While the literature on the relationship between intellectual property rights and innovation is vast, the literature on how open innovation relates to intellectual property and intellectual property policy is very limited. Intellectual property is generally accepted as a powerful asset that can be proactively

95 Zachary Quinlan, "Hindsight Bias in Patent Law: Comparing the USPTO and the EPO" (2013) 37(6) Fordham International Law Journal 1787; see also The Royal Society, *Keeping Science Open: The Effects of Intellectual Property Policy on the Conduct of Science* (2003) https://royalsociety.org/topics-policy/publications/2003/keeping-science-open/ accessed May 28, 2018.

managed, developed, and maintained to enhance business value.[96] However, there is also significant literature on how intellectual property rights conflict directly with the idea of open science and the free exchange of scientific knowledge.[97] Open innovation is perceived to promote unrestricted sharing of knowledge for the purposes of facilitating new discoveries. However, the literature has a tendency to confuse "open science" with "open innovation." Open innovation recognizes the value of knowledge exchange but does not promote the free sharing of knowledge at the expense of economic gain. In order to accelerate the introduction of new innovations into the market, open innovation must leverage the intellectual property system to facilitate access to and exchange of knowledge, to foster innovation in exchange for a degree of protection that will induce industry participation in the innovation process. For RRI and open innovation to achieve its objectives, patents can no longer be seen solely as an exclusionary and negative right. Previous studies have highlighted the potential negative impacts and risks of intellectual property rights in the context of biobanking.[98] However, the assertion of intellectual property rights does not necessarily mean that such rights will be exercised in an exclusionary manner. There is evidence that intellectual property rights are often not exercised in a manner that negatively impacts the research environment.[99] Intellectual property rights are required to stimulate and jumpstart the innovation process and to sustain businesses via the development of secondary markets for intellectual property.[100] It is therefore the role of RRI to arbitrate the degree of protection required in a given field of research, to balance between no intellectual property protection whatsoever (which would discourage risktaking and investment) and strong intellectual property protec-

---

96 Raymond Millien and Ron Laurie, "Meet the Middlemen" (2008) Intellectual Asset Management 28, 53–8; Henry Chesbrough and Roya Ghafele, "Open Innovation and Intellectual Property: A Two-Sided Market Perspective" in Henry Chesbrough, Wim Vanhaverbeke, and Joel West (eds) *New Frontiers In Open Innovation* (Oxford University Press 2014) 191–214.

97 See for example Bentwich, *supra* note 11; Subhashini Chandrasekharan and Robert Cook-Deegan, "Gene Patents and Personalized Medicine—What Lies Ahead?" (2009) Genome Medicine 1(9), 1–4; Pierre Azoulay, Waverly Ding, and Toby Stuart, "The Impact of Academic Patenting on the Rate, Quality and Direction of (Public) Research Output" (2009) Journal of Industrial Economics 57(4), 637–76; Hong and Walsh, *supra* note 11.

98 Edward S. Dove and Yann Joly, "The Contested Futures of Biobanks and Intellectual Property" (2012) Teoria Y Drecho: Revisita De Pensamiento Juridico 11, 132–46.

99 Timothy Caulfield, Robert M. Cook-Deegan, F. Scott Kieff, and John P. Walsh, "Evidence and Anecdotes: An Analysis of Human Gene Patenting Controversies" (2006) Nature Biotechnology 24(9) 1091–4.

100 *Ibid.*

tion (which would inhibit innovation). A balanced intellectual property policy that advocates for the interests of open innovation as well as the interests of all stakeholders involved will likely support follow-on innovations.

Open innovation therefore seeks to cobble together the efforts of various stakeholders involved in the innovation process and leverage existing resources to discover how best to apply and translate new knowledge into socially beneficial innovations. Because the research, development, translation, and commercialization of biobank-derived research are all distributed across multiple stakeholders, the innovation process must recognize and ensure each stakeholder receives the quid pro quo required to induce participation. Intellectual property therefore plays a role in supporting open innovation by enabling and promoting the exchange of knowledge to introduce innovations into the market for the benefit of the public. In the context of biomedical innovations, the assertion of patent rights is arguably necessary to incentivize and safeguard the interests of stakeholders that are motivated by commercial interests (industry, universities, and academic entrepreneurs) and to realize the objectives of stakeholders that are motivated by socioeconomic growth (government, the public, and donors). However, the commercialization strategies used to introduce new innovations into the market do not necessarily need to rely upon the traditional exclusionary right associated with the patent system. Specifically, commercialization strategies often employed by the biomedical industry, such as—but not limited to—patent pools, broad licensing, and reachthrough rights, have their foundations in asserting patent rights, but an innovation strategy that embraces patenting does not necessarily need to be exercised in an exclusionary manner that negatively impacts the advancement of scientific research. Open innovation should not be averse to the assertion of intellectual property but it should object to the irresponsible use of intellectual property rights. At its core, commercialization is not only about generating revenue; it is also about translating basic research into a form that the public can use and making publicly funded research available for the benefit of the public. "Openness" requires a willingness to question the process of knowledge production, translation, and commercialization to determine what systems are required and appropriate to society's current need for access to advancements in science and innovation.

5 HOLISTIC INNOVATION FRAMEWORK

EU policies on open access and open science make it very difficult for research institutions to adopt practices that respect the seemingly contradictory and competing objectives of innovation. On the one hand, there is literature that supports the assertion that intellectual property rights are a significant constraint on the advancement of scientific research because they create barriers to free exchange of scientific knowledge. On the other, universities urge researchers to protect the commercial potential of their research by patenting and forming close partnerships with industry to facilitate the translation of knowledge into products. Although the free exchange of knowledge seems inconsistent with intellectual property protection, in reality patent rights and open science are not necessarily irreconcilable. Both aim to maximize the impact of research and utility of scientific knowledge through full and enabling disclosure to foster follow-on innovations. After all, the foundation of the patent system rests on the exchange of full enabling disclosure for a period of exclusivity. In fact, the legal requirement of enabling disclosure may even make patents more open than academic publications, where researchers may withhold crucial information for personal reasons, such as preserving prospects of obtaining further research funding.[101] How patent rights are used may affect open access, but patents and the patent system are not inconsistent with the principles of open access, open science, and/or open innovation. For example, the recent trend of patent donations from industry to research institutions allows researchers to leverage existing (protected) knowledge and extend beyond it to facilitate research in other fields, to enable cross-industry innovations.[102] As such, commercialization and open access/open science could be seen as complementary strategies within a holistic innovation framework aimed at getting the optimal social and economic value from publicly funded research.

The apprehension and skepticism associated with commercializing innovations derived from biobank resources may be contributed to by the negative stigma associated with the word "commercialization."[103] The concept of money and profit in relation to using HBM as a capital resource is typically seen by society to be contrary to public policy.[104] However, commercialization

---

[101] Patrick Andreoli-Versbach and Frank Mueller-Langer, "Open Access to Data: An Ideal Professed but Not Practiced" (2014) Research Policy 43(9), 1621–33; Marie Thursby, Jerry G. Thursby, Carolin Haeussler, and Lin Jiang, "Do Academic Scientists Share Information with Their Colleagues? Not Necessarily' (2009) www.voxeu.org/article/why-don-t-academic-scientists-share-information-their-colleagues accessed June 4, 2018.

[102] Nicole Ziegler, Oliver Gassmann, and Sascha Friesike, "Why Do Firms Give Away Their Patents for Free?" (2014) World Patent Information 37, 19–25.

[103] *Supra* note 92; see also *supra* note 59, where the report found that the public expressed concern over profit motives and mistrust of commercial entities accessing biobank data despite having very little knowledge of the innovation process and role of industry in biomedical research.

[104] Council of Europe, The European Convention on Human Rights and Biomedicine (1997), Article 21 states, "The human body and its parts shall not, as such, give rise to financial gain". See also UNESCO, Universal Declaration on the Human Genome and

is a much more complicated word that incorporates the concept of making basic research available for the benefit of the public. Taking a pragmatic view and recognizing that only through translation and commercialization will new innovations for the betterment of human health be introduced may temper some of the concerns and criticisms. The focus should therefore be on making the best possible use of HBM for the benefit of patient interests, instead of a pure financial motive.[105] The reality is that advancing biomedical research requires the involvement of multiple stakeholders because none of the individual players in the innovation process have all the necessary skills and resources to research, develop, commercialize, and translate discoveries into innovations independently.[106] Furthermore, the public's desire for biobanks to share results and advance healthcare suggests that benefit sharing and/or return of benefit may help the public accept the notion of commercialization.[107] With the broader view of improving human health, the whole can be more than the sum of its parts if stakeholders can work together to effectively and efficiently translate publicly funded research into innovations that can benefit public health as well as generate funds for further academic research and drive socioeconomic growth.

Current scientific contributions are still fragmented and are far from presenting a holistic picture of open science and policy implications to address the issue of stakeholder incentives to advance the RRI agenda. There is some literature advocating "reconceptualizing" the commercialization conflict away from the argument of private versus public interests and focusing instead on knowledge production.[108] However, "reconceptualizing" a conflict is a theoretical exercise that does not directly acknowledge the interests or address the

---

Human Rights (1997), Article 4 which states "[t]he human genome in its natural state shall not give rise to financial gains".

105 Kathinka Evers, Joanna Forsberg, and Mats Hansson, "Commercialization of Biobanks" (2012) Biopreservation and Biobanking 10(1) 45–7.

106 *Supra* note 92.

107 Alexander M. Capron, Alexandre Mauron, Bernice S. Elger, Andrea Boggio, Agomoni Ganguli-Mitra, and Nikola Biller-Andorno "Ethical Norms and the International Governance of Genetic Databases and Biobanks: Findings from an International Study" (2009) Kennedy Institute of Ethics Journal 19(2), 101–24; Gillian Haddow, Graeme Laurie, Sarah Cunningham-Burley, and Kathryn G. Hunter, "Tackling Community Concerns About Commercialisation And Genetic Research: A Modest Interdisciplinary Proposal" (2007) Social Science & Medicine 64(2), 272–82.

108 Kean Birch, "Knowledge, Place, and Power: Geographies of Value in the Bioeconomy" (2012) New Genetics and Society 31(2), 183–201; Kean Birch and David Tyfield, "Theorizing the Bioeconomy: Biovalue, Biocapital, Bioeconomics or ... What?" (2013) Science, Technology & Human Values 38(3), 299–327; Andrew Turner, Clara Dallaire-Fortier, and Madeleine J. Murtagh, "Biobank Economics and

concerns of the public and donors. Asking the public and donors to simply "think differently" about commercialization does not appease or change public perceptions of profit motives and commercial agendas at the expense of public good. Current literature on open innovation and commercialization predominantly adopts a business-centric view, which continues to feed the public perception that the assertion of intellectual property rights is mainly motivated by profits.[109] Mentally navigating the perceived contradictions between private versus public interests and/or open versus closed innovation still requires the public to accept that a conflict exists. Instead of this, practical efforts should be made to engage with society, with the objective of helping the public understand that without the participation of all stakeholders, the advancement of biomedical research and the discovery of lifesaving therapies and treatment will not happen, or will stall if stakeholders choose not to play their role in the innovation process.

As such, incorporating the principles of RRI, a holistic innovation framework should focus on societal engagement to ensure that all contributors to the innovation process understand that translation is a complex multistakeholder process whereby each stakeholder plays an essential role in contributing to the successful development of basic research into socially beneficial outcomes. It is not enough for the stakeholders to merely accept their role in the innovation process. There must be a mechanism within a holistic innovation framework that (a) recognizes what motivates each of the relevant stakeholders *and* (b) enables the stakeholders to receive the specific benefit they expect in exchange for their contribution to the innovation process. To incentivize participation, every stakeholder must extract some reward or quid pro quo for doing so. By understanding that the translation process is a part of the innovation value chain, stakeholders can appreciate that they are each an essential part of a larger framework whereby their respective input is required to maximize the prospects of introducing a potentially lifesaving innovation to the benefit of the public.

A value chain can only work if all stakeholders play their respective role to achieve the common goal. To ensure participation, all stakeholders must be

---

the 'Commercialization Problem'" (2013) Spontaneous Generations: A Journal for the History and Philosophy of Science 7(1), 69–80.

[109] Edna Einsiedel and Lorraine Sheremeta, "Biobanks and the Challenges of Commercialization" in Christoph W. Sensen (ed.) *Handbook of Genome Research: Genomics, Proteomics, Metabolomics, Bioinformatics, Ethical and Legal Issues* (Wiley 2005), 537–59; Christine Critchley, Dianne Nicol, and Margaret Otlowski, "The Impact of Commercialisation and Genetic Data Sharing Arrangements on Public Trust and the Intention to Participate in Biobank Research" (2015) Public Health Genomics 18(3), 160–72.

able to extract their desired reward. In this context, biobank donors want to benefit from the research that uses their samples. The general public expects a return of benefit from the tax dollars the government uses to publicly fund basic research. To obtain those benefits, industry needs to translate and commercialize early stage research and discoveries derived from biobank samples or data. However, industry will only participate if there is a financial incentive to do so. In order for industry to have anything to translate, researchers need to have the means to conduct research, preferably in an environment conducive to knowledge creation and diffusion. Researchers want to have free and open access to knowledge and data to advance scientific discoveries and disseminate results. The emergence of "academic entrepreneurs"[110] who are motivated by economic interests in their research and "entrepreneurial academics"[111] who operate as knowledge brokers by leveraging external grants for market oriented research adds a layer of complexity to the innovation landscape. Nevertheless, in order for researchers to conduct research, universities and public research organizations need to attract funding. Technology transfer and public–private R&D collaborations fostered by universities close the gap between science and practice by merging competences to find applications that address societal challenges. When successful, the resulting research excellence and innovation brings prestige to academia and increases prospects of attracting funding.[112] Government will provide public funding for basic research so long as the funds go toward supporting research with the potential to drive economic growth and address societal challenges.[113] The socioeconomic growth sought by government will only be realized if industry can successfully commercialize basic research (that is, create jobs, generate tax revenue, and introduce innovations to the market to improve social welfare). Industry will also provide funds to support basic research if there is an incentive to invest, such as through intellectual property and commercialization policies that contemplate revenue sharing. The interconnectedness of stakeholder interests almost creates a chain reaction—one needs to happen in order for the next to follow.

---

[110] *Supra* note 86.

[111] Cris Shore and Laura McLauchlan, "'Third Mission' Activities, Commercialization and Academic Entrepreneurs" (2012) Social Anthropology 20(3), 267–86.

[112] See for example Bart Clarysse, Mike Wright, Johan Bruneel, and Aarti Mahajan, "Creating Value in Ecosystems: Crossing the Chasm between Knowledge and Business Ecosystems" (2014) Research Policy 43(7), 1164-117.

[113] See for example European Commission, "Innovation: How to Convert Research into Commercial Success Story? Part 3: Innovation Management for Practitioners" (2013) https://ec.europa.eu/research/industrial_technologies/pdf/how-to-convert-research-into-commercial-story_en.pdf accessed June 4, 2018.

The logistics of bringing a value chain together to develop an innovation is no small challenge but the key to optimizing the social and economic value of publicly funded biobanks and related biomedical research is effective and efficient collaboration between the public and private sector. If all the stakeholders involved in the innovation process understand the interconnectedness of their respective roles in the chain of events from discovery to translation and commercialization, and what is at stake if there is a disruption in the chain, there is at the very least a greater likelihood that basic research will be more efficiently translated into socially beneficial innovations.

The role of RRI principles should therefore be to govern the "responsible" participation of stakeholders in the translation and commercialization process, and not to shape the innovation process itself. RRI should not itself be the objective, but rather a way of organizing the interaction between stakeholders with the goal of facilitating the translation and commercialization of innovations and maintaining trust between the stakeholders so they are mutually responsive to each other and respectful of public values. Evidence indicates that the public has greater trust in public research institutions, even if an institution receives some private funding,[114] suggesting that publicly funded, nonprofit independent institutions involved in the translation of research may be the types of organization that are in the best position to maintain public trust and manage the translation and commercialization of research and discoveries derived from biobanks. Specialized institutions that attempt to bridge the gap between open science and open innovation in a particular field could work directly with specific industry sectors to explore new ways to integrate and apply university research to develop new innovations.[115] With their unique knowledge of the particular dynamics associated with the sector and with representation from each of the stakeholder groups, these independent translation organizations would essentially act as "guardians" and decide on how best to use patent rights to achieve the most desirable socioeconomic outcome. For example, a concrete way to manage the interaction between open science and intellectual property is to entrust the translation organizations with the right to define an access zone around the state of the art, and grant licenses to enable follow-on innovations when appropriate and allocate royalties derived

[114] Critchley et al, *supra* note 54.

[115] Other example of specialized institutions representing the interest of different fields include IMEC in Belgium, which specializes in combining basic research in microelectronics and nanoelectronics into semiconductor technologies: www.imec.org; and ATTRACT, a pan-EU initiative that specializes in accelerating the development of high-performance detector and imaging technologies through a process of co-innovation among European research institutes, small and medium enterprises (SMEs), companies, and universities: www.attract-eu.org.

from the patents to support the financial sustainability of biobanks. Different drivers and regulatory structures impact each industry sector and because translation organizations operate in the broad scientific community engaged in research and development activities of a particular sector, such organizations are arguably in the most neutral and informed position to determine the best combination of open science and intellectual property rights to derisk the translation and commercialization of research while ensuring that the most desirable socioeconomic outcome can be achieved. Patents do not necessarily create barriers or close the door to open science as there exists an acceptable balance in "the spectrum between free use of knowledge by anyone for any purpose, to exclusive use by one entity for its own use."[116] If the governance and regulation of such publicly funded research organizations is transparent and clearly communicates intentions and goals regarding how translation and commercialization achieve public health benefits, then the public may be more accepting of commercialization efforts.[117] This may be achieved through adopting guidelines regarding the management of intellectual property that incorporate RRI principles on governing the collaborative relationships and respective quid pro quo positions of the stakeholders.

6 CONCLUSION

The general public often voices opposition to the idea of commercialization in the field of biomedical research without truly understanding the process involved in translating discoveries into new innovations and making them safe and available for the benefit of the public. This lack of understanding is precisely what makes it so difficult for public policy and strategies to be implemented with respect to the commercialization of biomedical innovations derived from biobank samples and data. Resources need to be devoted to informing all stakeholders of the collective effort required to translate research into innovations. The public needs to understand that the process of bringing research and discoveries to market for the benefit of society requires the collaborative efforts of multiple stakeholders. Potential profit is what incentivizes industry participation. The economic growth sought by government will only be realized if industry can successfully commercialize innovations. Proprietary

---

[116] Dianne Nicol and Richard Gold, "Standards for Biobank Access and Intellectual Property" in Mattew Rimmer and Alison McLennan (eds) *Intellectual Property and Emerging Technologies: The New Biology* (Edward Elgar 2012), 133–57.

[117] Kristin S. Steinsbekk, Lars Ø. Ursin, John-Arne Skolbekken, and Berge Solberg, "We're Not in It for the Money—Lay People's Moral Intuitions on Commercial Use of 'Their' Biobank" (2013) Medicine, Health Care And Philosophy, 16(2) 151–62; Haddow et al, *supra* note 106.

rights in innovations are the foundation of building successful businesses, which leads to job creation, tax revenue, and socioeconomic growth. From the government's perspective, industry's success in translating and commercializing publicly funded research means the social and economic value derived from academic research has been realized, which in turn justifies further university and public research funding. Researchers and universities therefore benefit from the flow of private and public funding to conduct further basic research to advance science and meet humanitarian goals. Most importantly, both the public and individual donors benefit from the availability of therapies, medicines, and technologies derived from the generous donations of HBM in biobanks.

However, an equally plausible outcome of this holistic innovation framework is that the stakeholders may still fundamentally disagree with the price to be paid to incentivize other stakeholders' participation in order for innovations to be made available for the benefit of the public. Despite this possible lack of consensus, at the very least, efforts must be dedicated to engaging all the stakeholders in participation in the practicalities of the innovation process, in order to create a foundation for informed dialogue, if there is any chance at reconciling differences or finding a mutually agreeable solution. Informing stakeholders of the realities of the translation process in the value chain of innovation may make them more willing to collaborate and more accepting of compromises in order to achieve a common goal of social responsibility and public good. Reconceptualizing issues is not going to make people change their opinion, but making them see the interconnectedness of their respective roles in the translation value chain in order to achieve the ultimate objective of public benefit may help them recognize that the "quid pro quo" required to ensure the continued responsible participation of the other stakeholders is more valid. An overemphasis on individual interests without sufficient attention to the greater social, economic, and structural challenges of translation may undermine rather than protect societal interest.[118] As von Schomberg states, "RRI should be understood as a strategy of stakeholders to become mutually responsive to each other and anticipate research and innovation outcomes underpinning the 'grand challenges' of our time for which they share responsibility."[119]

[118] Lori Luther and Trudo Lemmens, "Human Genetic Data Banks: From Consent to Commercialization—An Overview of Current Concerns and Conundrums" (2012) Biotechnology 12, 183–217.

[119] Von Schomberg, *supra* note 24.

12. Do we need an expiration date for biobanks?

Franziska Vogl and Karine Sargsyan

1 INTRODUCTION

All over the world, biobanks are growing in number and size. In 2012 Boyer and colleagues identified more than 600 biobanks in the USA alone.[1] Biobanks take a pivotal role in advancing public health by revealing the aetiology of various diseases, by promoting innovation of diagnostics and treatment and by translation of innovation to bedside.

Those working in biobanking are convinced that biobanks have the potential to become the most powerful research infrastructure for health innovation—provided that they are sufficiently resourced and networked.[2] In this context, facilitation of use and efficiency play major roles. Maximizing the use, productivity, and value of biobanks worldwide will depend on a transition in the ways in which biobanking is perceived and conducted.[3]

Biospecimens have been collected for a long time. Originally, these collections were used to diagnose and treat a patient's disease, but the samples were also utilized for research purposes or to educate medical students.[4] However, biobanks have developed tremendously in the past decades, especially in terms of standardization and harmonization. Although biobanks are gaining increasing importance as research platforms in the field of biomedical research to provide access to large numbers of catalogued samples and/or data, most have not yet reached their full potential. Colledge and colleagues summed up the

---

1 GJ Boyer, W Whipple, RJ Cadigan, GE Henderson, "Biobanks in the United States: How to identify an undefined and rapidly evolving population" (2012) Biopreserv Biobank 10(6) 511–17.

2 JR Harris, P Burton, BM Knoppers, et al, "Toward a roadmap in global biobanking for health" (2012) Eur J Hum Genet. 20(11) 1105–11.

3 *Ibid.*

4 BR Jeffers, "Human biological materials in research: Ethical issues and the role of stewardship in minimizing research risks" (2001) ANS Adv Nurs Sci. 24(2) 32–46.

15 major barriers to sharing samples in a review article.[5] To overcome these barriers, prominent organizations have tried to develop regulations to open up biobanks for more researchers. Biobanks and biorepositories have gained practical knowledge from a major regulative paper by the European Commission with the title "European Charter for Access to Research Infrastructures: Principles and Guidelines for Access and Related Services,"[6] which sets out nonregulatory principles and guidelines to be used as a reference when defining access policies for research infrastructures and related services.[7]

The term "research infrastructure" takes in the facilities, resources, and services that are used by researchers and scientific communities (academic or industrial) to perform research and to facilitate research and innovation in their designated field. Infrastructure comprises the following: major scientific equipment and sets of instruments, knowledge based assets such as specific collections (including biological sample or compound collections), archives and scientific data of any type, and e-infrastructure such as data and computing systems and communication networks, as well as further infrastructure and tools that are essential to achieving excellence in research and innovation.[8]

Even though this definition of research infrastructure is well suited to a biobank, the policies for access to samples from a biorepository have more facets than the access policies for an analytical platform infrastructure. Some of the multiple facets of a biobank's access policy include ethical issues, legal issues, or availability of clinical and pathological data. The National Cancer Institute (NCI) provides best practice documents including the "NCI Best Practices for Biospecimen Resources,"[9] which outline the operational, technical, ethical, and legal best practices for NCI-supported biospecimen resources.

---

[5] F Colledge, B Elger, HC Howard, "A review of the barriers to sharing in biobanking" (2013) Biopreserv Biobank 11(6) 339–46.

[6] European Commission, "European charter for access to research infrastructures: Principles and guidelines for access and related services" (Publications office of the European Union, 2016) https://ec.europa.eu/research/infrastructures/pdf/2016_charterforaccessto-ris.pdf accessed April 25, 2018.

[7] *Ibid.*

[8] Council Regulation (EC) 1291/2013 on establishing Horizon 2020—the Framework Programme for Research and Innovation (2014-2020) and repealing Decision No 1982/2006/EC [2013] OJ L 347/104, article 2 (6).

[9] National Cancer Institute, "NCI Best Practices for Biospecimen Resources" (2016) https://biospecimens.cancer.gov/bestpractices/2016-NCIBestPractices.pdf accessed April 25, 2018.

According to this, biospecimen resources should establish guidelines for sample distribution and clinical data sharing with the following characteristics:

- *Clear* to ensure their comprehension and adoption;
- *Flexible* to allow application to diverse and evolving scientific needs
- *Amendable* to facilitate their adaptability over time.[10]

The focus of the research papers and regulations cited in this introductory section is on availability and accessibility of samples, regulations, sample quality, and the need for harmonization in the biobanking sector. None of these publications discuss a possible expiration date for the use of samples from biobanks.

In the late 1990s, researchers from the RAND consortium struggled to catalog all collections of banked tissue specimens in the United States. From the tissue collections described in this book, a conservative estimate is that more than 307 million specimens from more than 178 million cases were stored in the United States at the time.[11] Since the RAND study, tissue collections and biobanks have boomed in different settings and variations, further confusing any census or cataloging of the banked samples.

Even though a huge number of samples and data are stored in biorepositories all over the world, only a few ideas were developed about the expiration date of these collections. However, the expiration date of informed consent was discussed in research articles in many different counties. The publications addressing this topic also discuss the possible expiry date of informed consent and many state that informed consent can have—or even should have—an expiry date, although some authors argue against this.[12] The papers state that privacy is a crucial and essential value in liberal societies. The research subject as an individual may decide on his/her own about an end point and/or open-end consenting. If there is an end point to the informed consent, this implies that the samples covered by this consent must also have an expiration date.[13]

Aside from discussions about an expiry date of informed consent, there are only a few reports about expiration or termination of a biobank. Biobank Graz calculated the hypothetical costs of termination of one sample collection.[14] Surprisingly, the costs for termination were higher than the costs for continu-

---

10 *Ibid.*

11 E Eiseman, S Haga (eds), *Handbook of human tissue sources: A national resource of human tissue samples* (RAND Corporation 2000).

12 W Peissl, *Privacy—Ein Grundrecht mit Ablaufdatum?—Interdisziplinäre Beiträge zur Grundrechtsdebatte* (VÖAW 2003).

13 *Ibid.*

14 T Macheiner, B Huppertz, M Bayer, K Sargsyan, "Challenges and driving forces for business plans in biobanking" (2017) Biopreserv Biobank 15(2) 121–5.

ing the collection for the next 3–4 years. Colleagues from London published an account of their difficulties with biobank closure, keeping the biobank facility itself anonymous (pseudonymized in the paper as "Xbank").[15]

For this biobank, the reason for termination had to do with the business model: The funders felt that the biobank had not collected sufficient tissue samples in its first years and were of the opinion that it was not performing well enough. Furthermore, they decided that other biobanks in the same area had sufficient material.[16]

Termination due to a lack of funding is a typical scenario for a biobank. The example of "Xbank" shows that the expiry date of a biobank facility may be directly connected to the expiry date of funding.

2 DEFINING "EXPIRATION DATE"

In the context of biobanking, the term "expiration date" may be discussed from the perspective of sample quality as well as from an ethical and/or legal perspective. In the latter context, "expiration date" means that access to specific samples may be ended, that is, that the ethics committee vote has expired. Accordingly, the "expiration date of a collection" is the date on which accessibility ends or a collection comes to an end.

2.1 Collection Strategies

The requirements for biobank samples have changed enormously in the past 20 years, and with the emergence of new sample handling techniques and new technologies for sample analysis, sample quality has become more and more important. Moreover, the digitalization of health information has enabled the storage of large data sets. Quinlan and colleagues state that modern biobanking must change its strategic focus from a sample-dominated perspective to a datacentric strategy.[17] However, this change of paradigm also affects biobanks' collection strategies, which need to be adapted from time to time to ensure they remain up to date, since research interests change and technologies

---

15 N Stephens, R Dimond, "Unexpected tissue and the biobank that closed: An exploration of value and the momentariness of bio-objectification processes" (2015) Life Sci Soc Policy 11:14-015-0032-0.

16 N Stephens, R Dimond, "Closure of a human tissue biobank: Individual, institutional, and field expectations during cycles of promise and disappointment" (2015) New Genet Soc. 34(4) 417–36.

17 PR Quinlan, S Gardner, M Groves, R Emes, J Garibaldi, "A data-centric strategy for modern biobanking" (2015) Adv Exp Med Biol. 864 165–9.

develop over time. In what follows, three practical examples for this situation are illuminated in more details.

2.1.1 New methodology—old collection strategies come to an end

A typical example is a new tool referred to as "liquid biopsy." This method (plasma sequencing) enables the detection of early stages of cancer progression using cfDNA (cell free DNA) isolated from blood. The method is very susceptible to deviations in preanalytical sample handling: the plasma needs to be separated from the blood cells immediately after blood collection; centrifugation is a critical part of the sample preparation process; centrifugal speed must not be too high (to avoid cell lysis) and multiple centrifugation steps are recommended.[18] Plasma sequencing is one of many examples showing how researchers' sample quality requirements have changed with the emergence of new technologies. To meet these new criteria, old collection strategies come to an end and are replaced by new ones.

2.1.2 Collections with predefined expiry dates

Sample collection in the course of a clinical trial is an example for a collection with a predefined expiry date. After completion of the clinical trial, it is necessary to deposit the samples unless it is possible to use the samples outside the context of the clinical trial. In this case, the wording of the informed consent, the IRB decisions, and the legal framework determine whether the samples may be used for other research or not.

2.1.3 Diagnostic sample collections are often open-ended

Sample collections for diagnostic or therapeutic purposes (such as pathology archives) have predefined expiry dates. In Austria, no concrete legal obligation exists regarding the length of time over which samples collected for diagnostic purposes should be stored. For example, Biobank Graz stores these samples for a time period of 30 years.

Nevertheless, not all samples stored in biobanks have an expiration date. Certain collections, such as disease-focused or population-based collections, are open-ended. As long as sample and data quality are high and the samples are kept under adequate storage conditions, they can be stored for many centuries.

The term "expiration date" may also be used in the context of sample quality. But what is the lifespan of a sample stored in a biobank? From daily experience

---

[18] L Benesova, B Belsanova, S Suchanek, et al, "Mutation-based detection and monitoring of cell-free tumor DNA in peripheral blood of cancer patients" (2013) Anal Biochem. 433(2) 227–34.

at Biobank Graz, one of the largest repositories of clinical samples in Europe, we derived that the lifetime of biomaterials depends on five main factors (see Figure 12.1). These factors are: (1) collection strategy (just discussed); (2) sample quality; (3) storage conditions and duration and sample type; (4) availability of data; and finally (5) intended application. Thus, the "expiration date" of a sample is not one fixed date, but rather is flexible and strongly dependent on the intended application. In what follows, the five factors that largely determine the lifetime of a sample are discussed in greater detail.

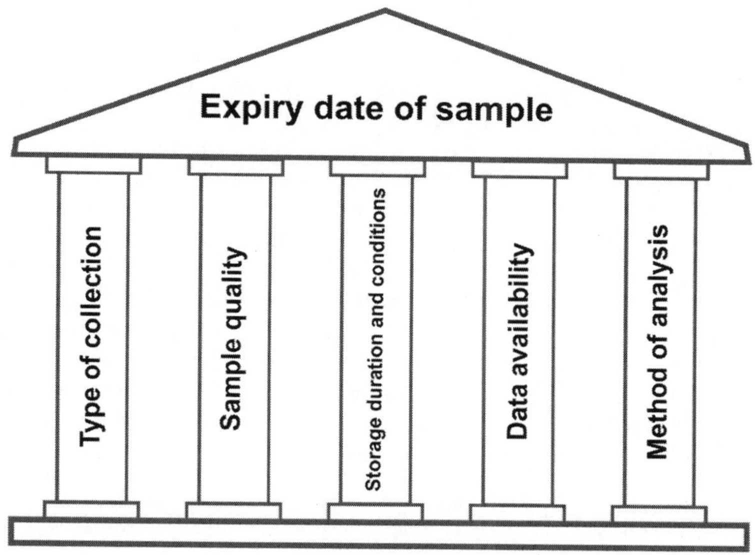

Figure 12.1 *Five pillars that determine the expiry date of a biospecimen that is stored in a biorepository*

2.2 Sample Quality

In the recent past, sample analysis technologies have become more sophisticated, and concomitantly sample quality has become increasingly important. Nevertheless, researchers complain about inadequate access to quality biospecimens.[19] Preanalytical variables, such as ischemia time, temperature,

[19] HA Massett, NL Atkinson, D Weber, et al, "Assessing the need for a standardized cancer HUman biobank (caHUB): Findings from a national survey with cancer researchers" (2011) J Natl Cancer Inst Monogr. (42) 8–15; D Simeon-Dubach, P

or type of preservation method have been shown to cause the greatest variability in quality of various biomolecules such as RNA, DNA, or protein.[20] For example, tissue ischemia time significantly affects gene and protein expression patterns within minutes following surgical tumor excision.[21]

2.3 Storage Conditions and Duration

Storage conditions, predominantly temperature and humidity, are critical factors for long-term preservation of biological materials. Different environmental storage conditions are required depending on the type of biospecimen. Storing samples under suboptimal conditions has detrimental effects on quality and thereby reduces the sample lifetime. As a general rule, samples stored at lower temperatures are better preserved.

The most common sample types in biobanking are FFPE (formalin-fixed paraffin-embedded tissue), FF (fresh frozen tissue), and biofluids (various types of blood samples, urine, cerebrospinal fluid, and so on). Most biobanks routinely store FFPE materials (blocks and slides) at room temperature. Studies show that DNA, RNA, and protein extracted from 12 year old blocks show no significant difference in comparison to current year blocks.[21] In contrast, biofluids should be stored at –20 °C or –80 °C. The gas phase of liquid nitrogen (–197 °C) is optimal for long-term storage of cryopreserved tissue samples. Hubel and colleagues summarized various studies that demonstrate the importance of storage temperature on the stability of critical biomarkers for fluid, cell, and tissue biospecimens. They provide a summary of the scientific

Watson, "Biobanking 3.0: Evidence based and customer focused biobanking" (2014) Clin Biochem. 47(4–5) 300–8.

20 NJ Caixeiro, K Lai, CS Lee, "Quality assessment and preservation of RNA from biobank tissue specimens: A systematic review" (2016) J Clin Pathol. 69(3) 260–5; P Ahmad-Nejad, A Duda, A Sucker, et al, "Assessing quality and functionality of DNA isolated from FFPE tissues through external quality assessment in tissue banks" (2015) Clin Chem Lab Med. 53(12) 1927–34; SH Hong, HA Baek, KY Jang, et al, "Effects of delay in the snap freezing of colorectal cancer tissues on the quality of DNA and RNA" (2010) J Korean Soc Coloproctol. 26(5) 316–23; JJ Lou, L Mirsadraei, DE Sanchez, et al, "A review of room temperature storage of biospecimen tissue and nucleic acids for anatomic pathology laboratories and biorepositories" (2014) Clin Biochem. 47(4–5) 267–73; A Spruessel, G Steimann, M Jung, et al, "Tissue ischemia time affects gene and protein expression patterns within minutes following surgical tumor excision" (2004) BioTechniques. 36(6) 1030–7.

21 TJ Kokkat, MS Patel, D McGarvey, VA LiVolsi, ZW Baloch, "Archived formalin-fixed paraffin-embedded (FFPE) blocks: A valuable underexploited resource for extraction of DNA, RNA, and protein" (2013) Biopreserv Biobank 11(2) 101–6.

literature relevant to the stability of specific biomarkers in human fluid, cell, and tissue for the range of temperatures between –20 °C and –196 °C.[22]

2.4 Sample Type

The "expiration date" of a biospecimen not only depends on its quality and the storage conditions, but also on the type of sample. Different sample types, such as wet tissue, frozen tissue, paraffin-embedded tissue, glass slides, blood, serum, and urine can be kept for different storage periods. As an example, native tissue can be stored at +4 °C only for a few days, whereas the same tissue sample may be stored for centuries after paraffin embedment.

2.5 Available Data

Availability and accessibility of high-quality clinical and pathological data also strongly influences the expiry date of a sample. In the context of biobanking, the general term "data" may be subdivided into two large subcategories: sample associated data and patient associated data. Sample associated data comprises storage and processing data, attributes such as diagnosis and histology for tissue samples, and laboratory findings (for example, blood test results for blood samples). Patient associated data includes clinical data as well as, for example, demographic, socioeconomic, or lifestyle data.

Ideally, the full clinical history of the donor, including data on therapy and response, follow-up, outcome, comorbidities, and so on will be available and stored in the biobank databases.

2.6 Intended Application

The biomolecules to be analyzed—for example, RNA, DNA, protein, or lipid—have a huge impact on the expiration date of the respective samples. Some biomolecules, such as DNA, are highly stable; others, such as RNA or proteins, are easily degradable by enzymes that are ubiquitously present (RNAses, proteases).

Which particular assays are planned with the samples? For measurement of enzymatic activity or cell viability the sample should be as fresh as possible, whereas morphology assessment still works with a 30 year old paraffin-embedded tissue on a glass slide.

---

[22] A Hubel, R Spindler, AP Skubitz, "Storage of human biospecimens: Selection of the optimal storage temperature" (2014) Biopreserv Biobank 12(3) 165–75.

What is the goal of the planned experiments? Is a quantitative or qualitative analysis planned? Protein or mRNA expression profiling are typical examples for quantification. In contrast, the detection of a specific gene mutation is qualitative, as long as no information about the levels of the mutation is involved. Generally, samples for quantification of biomolecules may be used within a shorter time frame than samples collected for qualitative research.

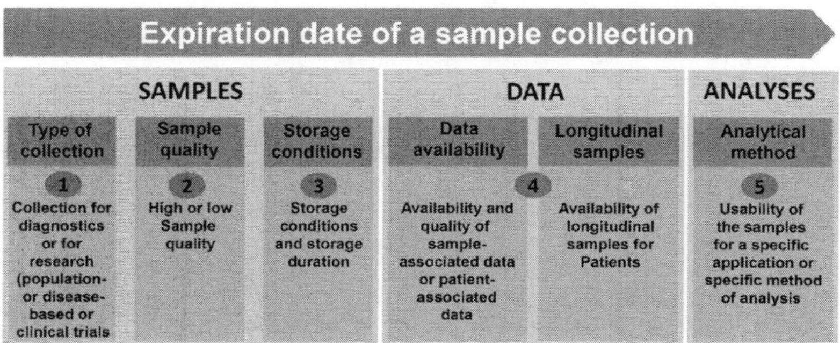

Figure 12.2 *Five main factors determine the expiry date of biospecimens stored in biorepositories: the type of collection, the sample quality, the storage conditions, the availability of data and the method of sample analysis*

2.7 Data vs Information

With the emerging omics technologies (proteomics, lipidomics, genomics, metabolomics, and so on), the volume of data that can be generated from biospecimens has grown rapidly over the past decades. The terms "data explosion" and "big data"[23] are used to describe this development. For example, the amount of genetic sequencing data stored at the European Bioinformatics Institute took less than a year to double in size.[24]

However, the generation and storage of a huge amount of data does not automatically translate into a huge amount of useful information. For data to become information, data processing is necessary: information can be generated by organizing, structuring, and interpreting data. This implies that

---

[23] A Gandomi, M Haider, "Beyond the hype: Big data concepts, methods, and analytics" (2015) International Journal of Information Management 35(2) 137–44.

[24] EMBL–European Bioinformatics Institute, "Annual Scientific Report 2012" (EMBL-EBI, 2013) https://issuu.com/embl/docs/embl_ebi_asr_2012_lo-rez accessed 25 April 2018.

valuable information can only be generated from meaningful data ("Garbage in, garbage out"). In biobanking as well as in other disciplines dealing with "big data," it is obligatory to think about the removal of old data that is never used ("junk data"). In the course of this, a trade-off is necessary between the future value expected from stored data and the costs of data storage, data management, and data processing. This consideration leads to the next section, which considers the comparison of the value of a sample with its costs.

3 VALUE ANALYSIS IN BIOBANKING: VALUE VS COSTS OF BIOSPECIMENS

In the previous section, the five main factors that determine the lifetime of biomaterials were discussed in detail: (1) collection strategy; (2) sample quality; (3) storage conditions, storage duration, and sample type; (4) availability of data; and finally (5) the intended application. Besides these five factors that focus on sample and data quality, an economic point of view is also relevant. To examine the quality of a product in comparison to the costs involved, value analysis is a useful management tool. For a biobank, the "products" are samples and data. In simple words, the general goal of a value analysis is to meet customers' demands at the lowest possible cost.

The value of a sample stored in a biorepository is determined by its actual usability for research (see the abovementioned five main factors). But in terms of costs, most biobanks do not have fixed prices for samples and data, in contrast to commercial tissue vendors. This is due to the complex tasks involved in the selection of patients and samples and in acquisition of clinical data. On the one hand, compared to other sources of biological specimens, clinical biobanks have access to a tremendous amount of patient clinical data. On the other, the search for specific samples in this large pool of samples and data is time-consuming. Such a query involves accessing various databases, different subcollections, and multiple departments. Consequently, the major cost factors for biobank services are the costs of sample collection, processing, and analysis; the costs of data management; the costs of pathology view; and finally the basic costs (comprising costs related to personnel, infrastructure, and maintenance).

Figure 12.3 shows the value of samples (blue color) in relation to the costs (red color) over time. The value of high-quality samples increases over time through their enrichment with new (follow-up) data and the collection of multiple longitudinal samples from the respective patients. This increase in value over time can be termed an "enrichment effect."

In contrast, the value of low-quality samples (no standardized collection, processing, and storage) will decrease over time in storage, since sample quality was already poor at the very beginning. In the latter case, sample handling in the preanalytical phase (sample collection and processing) mainly determines the poor sample quality.

The third possibility shown in Figure 12.3 is high-quality samples with a small but useful data set (minimum data set). These samples will neither increase nor decrease in value and will stay more or less constant over time.

Different biobanks have diverse definitions of the minimum data set. At Biobank Graz, the minimal data set comprises the following data: age, gender, date of death, pathological diagnosis, sample type, and data on storage and processing. BBMRI developed a minimum data set for biobanks and studies using human biospecimens. The data set is called MIABIS (Minimum Information About Biobank data Sharing).[25]

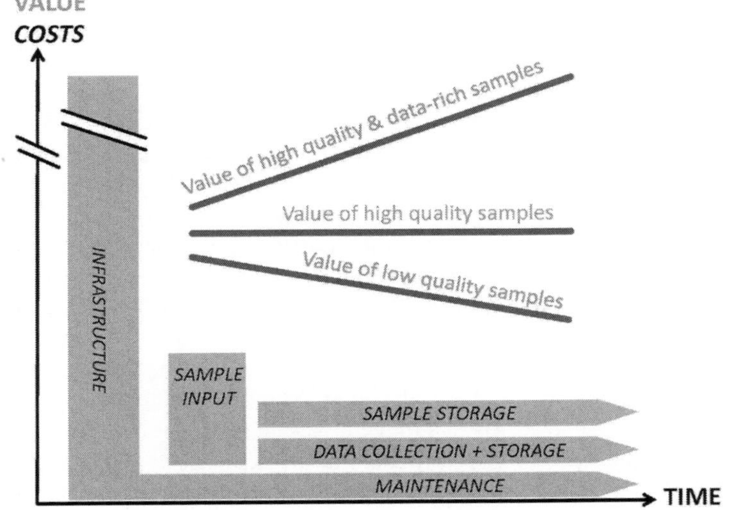

Notes: The value development of biospecimens over time is directly linked to sample quality and amount and quality of associated data. High-quality, data-rich samples increase in value over storage time. The value of high-quality samples with a minimum data set is expected to remain stable over a period of years. In contrast, low-quality samples further decline in value.

Figure 12.3 *Value of samples (blue color) in relation to the costs (red color)*

---

[25] L Norlin, MN Fransson, M Eriksson, et al, "A minimum data set for sharing biobank samples, information, and data: MIABIS" (2012) Biopreserv Biobank 10(4) 343–8.

4 TERMINATION COST MODEL FOR BIOREPOSITORIES

As discussed previously, no or only very few biobanks have included termination concepts or exit strategies in their future plans. Academic biobanks in particular are not willing to think about termination, and most academic biobanks have not found it necessary to terminate any such activity.

When planning an exit strategy workflow and/or calculating the costs for such a scenario, the following important points need to be taken into account:[10]

- biological samples require professional recycling (specific regulations are dependent on the local law)
- deconstruction and dismantling of infrastructure is important and can lead to higher costs (also for recycling of the material)
- for human biobanks the maintenance of paper archives can be mandatory, independent of biobank termination (depending on the law and regulations of the country). For example, in most European countries informed consent given should be archived for at least ten years after the end of the research activities.

Sargsyan and colleagues have shown that the costs of terminating the FFPE sample collection of Biobank Graz in a way that conforms with the relevant laws would be rather high.[26] Should such a termination be required in the future (for whatever reasons), the amount of money required to meet these costs would need to be available at that time.

5 SUMMARY

Only a few publications on the expiration date of biological sample collections are available. In contrast, the expiration date of informed consent has been discussed in multiple research articles.

This chapter discussed the term "expiration date" in the context of biobanking from various different perspectives: financial, ethical, and legal. In addition, we elucidated the impact of sample quality, data availability, and analytical methods on the lifespan of biospecimens.

From daily experience at Biobank Graz, one of the largest repositories of clinical samples in Europe, we derived that the lifetime of biomaterials depends on five main factors: (1) collection strategy; (2) sample quality; (3) storage conditions and duration and sample type; (4) availability of data; and finally (5) the intended application. Thus, the "expiration date" of a sample is

---

[26] *Supra* at note 9.

not one fixed date, but rather is flexible and strongly dependent on the intended application and on the financial resources of the respective biobank.

The most probable reason for a biobank's termination is financial issues (in most cases a lack of funding). Access to long-term funding for biobanks is still a problem and strategies to recover biobanking costs are emerging. The use of a well-functioning and expensive infrastructure for only one collection strategy is unusual. A considerable amount of biobanks are focusing on diversity and running both disease-based and population-based collections. Usually the collections for retrospective and epidemiological studies are not time limited. Therefore research usage of biospecimens is unspecified in terms of time and matter.

Before now, a biobank's expiration date has never been a point of discussion. In this context, the question of how much it costs to terminate a biobank is a peculiar one. The calculation made at Biobank Graz should be considered as a starting point in future discussions.

13. Biobanks and biobank networks

Eva Ortega-Paíno and Aaro Tupasela

1 INTRODUCTION

Even now, in almost the third decade of the twenty-first century, one of the remaining and major scientific challenges to be overcome is the lack of reproducibility in medical research. In mid-2015, Freedman and others published an article in PLOS Biology showing low reproducibility rates within life science studies.[1] The authors highlighted that half of the annual research investment in the USA is lost due to irreproducibility. This irreproducibility, caused mainly by preclinical issues such as study design, data analysis, laboratory protocols, and inappropriate biological reagents and reference materials, points to a lack of a standards and best practice frameworks. Furthermore, this continuous lack of standardization—for example, in the case of biological material, not only in collection but also in storage—could lead to the discovery of false biomarkers. In his comment published in *Nature*, Poste highlighted the huge gap existing between the literature, where more than 150,000 articles have been published claiming thousands of potential biomarkers, and the reality, where only around a hundred are used in clinics.[2] Therefore, there is an increased need for more stringently characterized biospecimens in biomedical research.

In the same article, Poste noted the need to demand from funding agencies—such as the NIH and the European Framework Programme, among others—better returns for the money invested in biomarker research. To achieve these goals, it is important to establish common standards, not only for storage and processing of biospecimens but also for discovery and validation, as well as harmonized data from those samples. As if this were not enough, it should also be compulsory to possess the whole range of cross-disciplinary capabilities needed for translational research, that is to say, from laboratory to the clinics.

---

1 L.P. Freedman, I.M. Cockburn and T.S. Simcoe, "The Economics of Reproducibility in Preclinical Research" (2015) 13 PLOS Biology 1.

2 G. Poste, "Bring on the Biomarkers" (2011) 469 Nature 156.

Therefore, partnerships between academic laboratories, industry, and health care should be encouraged in order to form larger research networks, and the need for high-quality samples should be prioritized.

In an era of high throughput technologies and the search for new potential biomarkers for therapy, prognosis, or even diagnosis, it is of great importance to gather samples from donors where the *preanalytical variables* have been minimized as far as is possible. These preanalytical variables are classified into a first group called *in vitro* (those variables that happen in the process from "needle to freezer," such as haemolysis, delay in centrifugation, transport, and so on) and a second group known as *in vivo*, such as drugs, alcohol, or cigarette consumption, as well as other factors such as sex, weight, age, habitat, and life-style. All these variables, *in vivo* and *in vitro*, should be controlled, annotated, and taken into account while selecting individuals for a study, with the aim of matching as many of them as possible to reduce interferences in the study and, therefore, the ultimate analysis.

So, where can these very high-quality specimens be found? The answer is simple: from harmonized and standardized biobanks. *Time Magazine* listed biobanks as "one of the ten ideas that will change the world."[3] According to Hewitt and Watson, biobanks are collections of human, animal, plant, and microbial samples, and the data associated to them, that must be managed according to professional standards.[4] The most common biobanks contain human material such as whole blood, plasma, serum, urine, and sperm, among others. According to Asslaber and Zatloukal, there are many different types of biobanks; the most important ones for research, however, are population-based biobanks and disease-oriented biobanks.[5]

The fast development of the "-omics" platforms, such as genomics, transcriptomics, metabolomics, proteomics, and so on, has produced an enormous amount of genotypic and phenotypic data stored in several databases around the world, with samples associated to them spread across various biorepositories. Without a proper understanding of the conditions under which the samples were processed, it is unclear to what extent this data is meaningful in a clinical sense. The process of biobanking, therefore, needs better quality control from the very beginning of the biobanking process to the very end, when data is being analyzed.

---

3 A. Park, "10 Ideas Changing the World Right Now: Biobanks" (*Time Magazine*, 12 March 2009) 8.

4 R. Hewitt and P. Watson, "Defining Biobank" (2013) 11 Biopreservation and Biobanking 309.

5 M. Asslaber and K. Zatloukal, "Biobanks: Transnational, European and Global Networks" (2007) 6 Briefings in Functional Genomics and Proteomics 193.

As we mentioned, there are many different types of biobanks. Longitudinal population-based biobanks are undoubtedly the most common. Frank has noted that the population biobanks in the Nordic countries represent a type of collection where the entire population is a cohort.[6] The organization of the Nordic countries, where each citizen gets a personal number at the moment of birth, makes them an acknowledged "goldmine" for biobanking. Citizens' personal numbers or identification numbers are used to link information from each individual to all National Quality registers as well as to healthcare records. This provides researchers with unparalleled information about the state of health of an entire country and allows them to develop predictive disease models. These predictive approaches can be used as prevention in medical programs, since volunteers—with or without any disease—donate samples at different time points with the aim of finding biomarkers/genes that can explain different diseases, mainly caused by environmental factors but also by, for example, diet. The typical samples stored for this purpose are isolated DNA and whole blood, as well as the data associated to them (environmental factors, family history, lifestyle, and so on).

A notable example of this is the VIP (Västerbotten Intervention Program), which is a long-term project in the northern part of Sweden which invites individuals aged 40, 50, and 60 for screening. This study gathers in total around 254,000 inhabitants. The participants are asked to complete a questionnaire concerning various lifestyle factors and to donate a blood sample to be frozen for further research purposes.[7] In Denmark, the Danish Blood Donor Study (DBDS) has collected samples from more than 60,000 Danes and seeks to become the largest collection in Denmark.[8] Another example, in Skåne, the southern part of Sweden, is the MDC (Malmö Diet and Cancer study). This study started in the early 1990s as a screening survey for the middle-aged population of Malmö. The study gathered 28,000 participants, all from Malmö. As in the VIP study, the individuals had to answer a survey as well as donating blood. Of the participants, 62 percent were women. Cardiovascular risk factors were measured in a random subsample (n=6,000). Sixteen years after the first recruiting process, a new clinical examination and blood sampling was performed (n=3,700).[9]

---

[6] L. Frank, "Epidemiology: When an Entire Country Is a Cohort" (2000) 287 Science 2398.

[7] National Cancer Institute, "Northern Sweden Health and Disease Study" https://epi.grants.cancer.gov/Consortia/members/northsweden.html accessed June 4, 2018.

[8] The Danish Blood Donor Study, "Information about DBDS" www.dbds.dk/ginformationUK.htm accessed June 4, 2018.

[9] SND, "Malmö Diet Cancer" https://snd.gu.se/en/catalogue/study/ext0012 accessed June 4, 2018.

Population-based biobanks are a good complement to disease-based biobanks, in which samples from individuals have been stored after diagnosis and/or even treatment and are used to answer some specific disease-driven questions. Unlike population-based biobanks, disease-oriented biobanks can be used soon after the collection of samples has concluded, and have a lower associated cost. An example of this type of study, and the biobank associated to it, is the SCAN-B (Swedish Cancerome Analysis Network—Breast) project in the region of Scania (Sweden). This multicenter initiative started in 2010, enrolling hospitals from the southern healthcare region (Malmö, Lund, Helsingborg, Växjö, Halmstad, Kristianstad, and Karlskrona). Three years later, the region of Uppsala joined SCAN-B as well. The project's aim is to find new methods of diagnosis, prognosis, and treatment.[10]

Another example of a disease-based biobank is the Auria Biobank in Finland (www.auria.fi). Established in 2012, the biobank collects samples from patients coming into the region's hospitals after they have given consent. In 2014 the biobank also transferred a large number of samples from the University of Turku Hospital pathology collection, in accordance with the Finnish Biobank legislation, which was established in 2013.[11] What makes these types of biobanks unique and important is that they are able to capture a large number of samples from specific hospital regions and track patient outcomes over long periods. It is hoped that biobank configurations of this type will also provide industry with new resources for conducting research. [12]

These examples of the different biobank formats not only show the need to collect biospecimens to find an answer to why some people develop diseases and others do not, but also point to the need to develop common standards between the biobanks to ensure that samples are of similar quality, as well as to make sure that the results from one study are comparable to those from others. Furthermore, in relation to rarer conditions, it is difficult to find the necessary samples in just one biobank. This means that researchers may need to request samples from numerous biobanks in order to ensure sufficient statistical signif-

---

[10] SCAN-B, "Swedish Cancerome Analysis Network—Breast" http://scan.bmc.lu.se/index.php/Main_Page accessed June 4, 2018.

[11] A. Tupasela and S. Liede, "State Responsibility and Accountability in Managing Big Data in Biobank Research: Tensions and Challenges in the Right of Access to Data" in Brent Daniel Mittelstadt and Luciano Floridi (eds), *The Ethics of Biomedical Big Data*, vol 29 (Springer International Publishing 2016); A. Tupasela, "Tensions Between Policy and Practice in Finnish Biobank Legislation" (2015) 13 Biopreservation and Biobanking 379.

[12] H. Lehtimäki and others, "Commercialization of Biobank Data: The Case of Auria Biobank" in G.D. Sardana and T. Thatchenkery (eds), *Organization Development through Strategic Management* (Bloomsbury Publishing India 2017).

icance. Ensuring that samples are of a similar quality and standard means their comparison and use generates fewer errors.

2 THE RISE AND DEVELOPMENT OF BIOBANK NETWORKS

In the past few decades biobank networks have become an important framework and activity through which the hundreds of biobanks around the world coordinate and develop their activities. According to the Biobanking and Biomolecular Resources Research Infrastructure (BBMRI), in Europe alone, there are more than 500 biobanks, which contain more than 60,000,000 samples.[13] With such a large number of collections and samples, the need for common standards and procedures is of utmost importance in order to improve the quality and reliability of research results.

There is a clear need for collaboration in biobanking and the development of biobank networks serves as an important medium through which collaboration is facilitated. It is easy to understand that not only cancer but also other diseases are present worldwide, implying the need to compare different ethnic groups and subpopulations, patterns of disease, and prevalence in different areas in the world. Furthermore, some commentators have noted that it is difficult to publish findings of genetic association studies unless several populations have been combined, preferably from different countries.[14] This increased pressure to generate replicable results across populations has also contributed to an increased demand for specified standards among biobanks. Taking all this into consideration, it is important to join forces and gather samples from as many biobanks in the region, in the country, and across the world as possible.

During the past two decades, different international networking initiatives in Europe, North America, Africa, and Australia, such as P3G, CTRNet, TuBaFrost, BBMRI-ERIC, and the Australasian Biospecimen Network, among others, have contributed to the proliferation of biobanking networks. These actors are central in the collection, processing, storing, and dissemination of biological samples and the clinical data associated to them and are, therefore, the pillars and the key of success of well-organized and managed

13 P. Holub and others, "BBMRI-ERIC Directory: 515 Biobanks with Over 60 Million Biological Samples" (2016) Biopreservation and Biobanking 14, 559–62.

14 J. Kere, "Miten Suomessa kerättyjä DNA- ja kudosnäytteitä voidaan hyödyntää? [How Can DNA and Tissue Samples Collected in Finland Be Utilized]" 123 Duodecim 864.

biobanks.[15] Well-functioning network activities will be reflected in subsequent high-quality research.

Biobank networks are not only a way to provide samples and data to researchers. They also serve important social and normative functions. As Mitchell and Waldby have noted, the collection and exchange of human tissue samples is not just a technical but also a social act.[16] The processes of collecting, storing, and sharing samples constitutes, according to those authors, a "tissue economy," which serves as a conduit through which values and norms are also established and shared. Within biobanking networks there are a plethora of examples through which social norms and standards are established. These include forms of informed consent,[17] establishment of norms regarding data-sharing practices,[18] discussions surrounding appropriate pricing,[19] and quality standards for the collection and storage of samples.[20]

Biobank networks also play an important role in policy setting, both nationally and internationally. The need to organize biobanking activities nationally is in part related to the international standards and requirements set forth within international networks. In Finland, for example, the national biobanks have sought to organize themselves using the hub and spoke model (www.bbmri.fi/). The same organization principle can also be seen in other countries, such as Sweden and Norway, for example. The coordinating bodies of biobank networks also play an important role in organizing and developing common resources for network members. This can include training and education, as well as serve in facilitating the setting of common standards. Biobank networks also provide an opportunity for biobanks to gain access to new resources, such as financial opportunities gained through entering into research collaborations with new partners and collaborators. In this sense, biobank networks can serve as accelerators of research and provide important new avenues through which scarce and expensive resources are better used and maintained in the end.

---

15 J. Vaught, A. Kelly, and R. Hewitt, "A Review of International Biobanks and Networks: Success Factors and Key Benchmarks" (2009) 7 Biopreservation and Biobanking 143.

16 C. Waldby and R. Mitchell, *Tissue Economies: Blood, Organs, and Cell Lines in Late Capitalism* (Duke University Press 2006).

17 K. Hoeyer and L.F. Hogle, "Informed Consent: The Politics of Intent and Practice in Medical Research Ethics" (2014) 43 Annual Review of Anthropology 347.

18 J. Kaye and others, "Data Sharing in Genomics—Re-Shaping Scientific Practice" (2009) 10 Nature Reviews Genetics 331.

19 H. Odeh and others, "The Biobank Economic Modeling Tool (BEMT): Online Financial Planning to Facilitate Biobank Sustainability" (2015) 13 Biopreservation and Biobanking 421.

20 A. Carter and F. Betsou, "Quality Assurance in Cancer Biobanking" (2011) 9 Biopreservation and Biobanking 157.

3 BIOBANKS: A MATTER OF QUANTITY AND QUALITY

It is important to stress that quality in biospecimens (samples gathered in a harmonized, standardized manner as well as the data associated to them) is a compulsory requirement, but this is not enough for the development of high-quality projects. For this purpose, it is necessary to have a sufficient number of samples that can provide statistical power to the analysis. Therefore, quantity is also a key factor to consider. For many noncommunicable diseases (cardiovascular diseases, cancer, chronic respiratory diseases, diabetes),[21] a sufficient number of individuals are enrolled in research projects within the country and/or even the region. However, when talking about rare diseases the issue of quantity is relevant, since the number of cases is very low in each country and statistical power is a must in research. Nor can we forget that bio-pharmaceutical companies operate in a global world, and ethnic diversity must also be addressed. This need opens up the issue of international collaboration for research, which makes biobank networks of great significance. Even for common diseases, there is a need for larger sample sizes. Despite large prospective cohorts (LPC) having thousands of samples, one cohort may not have enough samples of a specific type to reach statistical significance. Therefore, biobank networks serve an important function of providing researchers the ability to search for specific sample types of specific patients in a broad range of biobanks from around the world.

These valuable collections need to be accessible, managed in a proper manner, and standardized, as has been highlighted previously in this chapter. The information about these samples should present a common ontology so researchers can exchange not only the samples but also the data associated to them. Different approaches have been taken for this purpose. One—MIABIS (Minimum Information About BIobank Data Sharing), launched within the BBMRI network—had as its goal the harmonization of information from biobank samples throughout Europe.[22] The goal of MIABIS was to develop a minimum data set, consisting of 52 attributes, for biobanks and studies in human biospecimens. This data, at the aggregate level, would allow the reutilization of samples and furthermore will save money and resources in the long run, at the same time increasing the quality of research through an increase in the number of specimens and associated data.

---

21 WHO, "10 Facts on Noncommunicable Diseases" www.who.int/features/factfiles/noncommunicable_diseases/en/ accessed June 4, 2018.

22 L. Norlin and others, "A Minimum Data Set for Sharing Biobank Samples, Information, and Data: MIABIS" (2012) 10 Biopreservation and Biobanking 343.

Although collaboration and exchange are key concepts for the success of any biobank network, as is data sharing, it is also important to remark that this data belongs to a number of individuals, who must have their integrity protected. For this reason, some questions have been raised concerning legal, ethical, and societal issues (ELSI) in biobanking. It is important to discuss these individuals' rights and protect their integrity.

In the next section, we will briefly describe the BBMRI-ERIC infrastructure in Europe, which has become one of the largest biobanking networks in the world. The network highlights the possibilities that formalized biobanking networks offer to members in relation to coordination of activities, as well as the development and harmonization of common standards.

4 BBMRI-ERIC AS A PLATFORM FOR BIOBANK NETWORKS

BBMRI-ERIC is a European research infrastructure which was first initiated in 2008, when it entered the European Research Infrastructure Preparatory Phase of the ESFRI roadmap. Between 2008 and 2011 its main funding came through the European Framework Programme 7. Five years later, in 2013, BBMRI-ERIC was awarded legal status; since then it has grown to encompass 16 full member states (Austria, Belgium, Czech Republic, Estonia, Finland, France, Germany, Greece, Italy, Latvia, Malta, the Netherlands, Norway, Poland, Sweden, and the United Kingdom), three observers (Cyprus, Switzerland, Turkey), and one international organization (IARC/WHO). However, countries such as the Netherlands and Sweden were pioneers in the organization and had already initiated linkage of their BBMRI activities in 2009. The research infrastructure offered by BBMRI-ERIC is important: it is designed as a research infrastructure whose goal is to support interaction between members, as well as coordinate activities. The ERIC status facilitates the collaboration between biobanks and biomolecular resources not only by providing access to their collections, but also by providing expertise and services free of charge. As such, it is not a funding organization, providing money for research, but rather serves to facilitate activities, which utilize samples and data from biobanks around Europe and elsewhere.

From a historical perspective, the BBMRI network sought to improve the interoperability of biobanks in Europe. As it notes in one of its early publications:

> The move towards a universal information infrastructure for biobanking in Europe is directly connected to the issues of semantic interoperability through standardized

message formats and controlled terminologies. The information infrastructure has become a critical component in life sciences research.[23]

This perspective highlights one of the major visions driving biobank networks, namely the possibility of combining both samples and data from biobanks across Europe, and ultimately the world. The concept of interoperability of biobanks and data derived from biobanks had its roots in early attempts at analyzing data from different population studies in Europe. The GenomEUtwin project was one of the first such attempts at developing a platform in which samples from more than 600,000 twins could be analyzed. In order to make the analysis possible, the groups had to develop common standards and terms to describe the different variables they had used, such as BMI.[24] From the experiences gathered by the GenomeEUtwin project and others, it became apparent that in order to further facilitate the interoperability of other types of biobanks, there needed to be a common "language" that could be used to facilitate the comparison and use of samples and data. This "comparability" was one of the main tasks tackled in the Global Biobank Week, held in September 2017 in Stockholm. The need for harmonization within the different biobanks around the world was extensively discussed. This harmonization should be taken into consideration in the different steps from needle to freezer and back to the patient, and should encompass not only the sample but also the data associated to it.[25]

The development of the BBMRI research infrastructure took place within the context of the European research infrastructure (ERA), which has been developed, among other things, with the Innovation Union concept in mind.[26] The development of biobanks as a common resource for research, however, has not been the mission of the European Union alone. The OECD has also had an interest in promoting the development and use of biobanks to bolster research activities, as well as innovation.[27] In addition to bolstering biomedical

---

23 BBMRI (2007) Biobanking and Biomolecular Resources Research Infrastructure. Construction of New Infrastructures—Preparatory Phase. FP7-INFRASTRUCTURES-2007-1 research proposal, 29.

24 J. Muilu, L. Peltonen, and J.-E. Litton, "The Federated Database—a Basis for Biobank-Based Post-Genome Studies, Integrating Phenome and Genome Data from 600,000 Twin Pairs in Europe" (2007) 15 European Journal of Human Genetics 718.

25 M.K. Henderson and others, "Global Biobank Week: Toward Harmonization in Biobanking" (2017) 15 Biopreservation and Biobanking 491.

26 M. Mayrhofer, P. Holub, A. Wutte, and J.E. Litton, "BBMRI-ERIC: The Novel Gateway to Biobanks" (2016) 59 Bundesgesundheitsbl, 379–84.

27 OECD, "OECD Best Practice Guidelines for Biological Resource Centres" www.oecd.org/sti/biotech/38777417.pdf accessed June 4, 2018.

research in universities, the hope of these initiatives and efforts has been to encourage the use of biobanks in industry.[28]

Within the BBMRI network, each member state is expected to sign a memorandum of understanding (MoU) in which they agree to fund the development of biobanking activities and facilities within their own countries. This commitment is important in that it is focused on the support for structures that maintain biobanking facilities and activities. At the same time, member states are expected to fund research which utilizes biobanking samples. Given the vast differences in the number and scope of biobanks in various countries, the ways in which each member state organizes its activities is left up to them. A common example of this system is the hub and spoke model, where, within a country, there will be one central coordinator which supports the various biobanks around the country.[29]

BBMRI-ERIC also developed the idea of the Expert Centre (EC). An EC, following international standardized conditions, should be the link between public and private sector organizations carrying out the analysis of biological specimens. These EC, according to van Ommen, GJB, and others, will combine the expertise of both academia and industry, and this alliance will lead to a win–win scenario in which: (1) collaborations in research will increase; (2) information, knowledge, data, and technologies will be shared between partners; (3) ethical and legal issues will be dealt with correctly under the frame of the adequate code of conduct; (4) resources will be managed and used in a more efficient manner; (5) questions regarding ownership (of samples and data) will be minimized; and finally (6) an innovation will be the outcome of competitive research and development within the partners.[30]

There are several EC that could be used as examples. Among them, BARCdb, a platform developed in Sweden (www.barcdb.org), links biobanks to molecular technology platforms in the worldwide "-omics" scenario.[31] Other examples include the EGC (Estonian Genome Centre) and SciLifeLab (Science for Life Laboratory), a state-of-the-art national initiative of largescale molecular analyses. SciLifeLab offers services to both academia and industry

---

[28] G.-J.B. van Ommen and others, "BBMRI-ERIC as a Resource for Pharmaceutical and Life Science Industries: The Development of Biobank-Based Expert Centres" (2015) 23 European Journal of Human Genetics 893.

[29] M. Bruinenberg and others, "Comparing the Hub-and-Spoke Model Practices of the LifeLines Study in the Netherlands and the H3Africa Initiative" (2014) 12 Biopreservation and Biobanking 13.

[30] *Supra* note 26.

[31] J. Galli and others, "The Biobanking Analysis Resource Catalogue (BARCdb): A New Research Tool for the Analysis of Biobank Samples" (2015) 43 Nucleic Acids Research D1158.

and, in addition to new developed technologies, provides bioinformatics support.

In the previous section, we discussed the MIABIS system, which has been designed as a system setting out the minimum amount of information that should be available regarding biobank samples. From a broader perspective MIABIS plays a crucial role in the functioning and role of biobank networks. From a political and social perspective, the MIABIS system and its development and implementation can be seen as a technical translation and implementation of European research policy in which human tissue samples and their related information become a system through which European integration can be achieved on one level.[32] In this sense, biobank networks serve a broader political role in unifying European research practices. The MIABIS system serves as the basis for the smoother and more efficient functioning of the European tissue economy in relation to the biobanks, which agree to meet these standards. It helps to create and facilitate a system of exchange through the establishment of common standards.

The network as a metaphor for activity across borders is also emblematic of the emergent governance structure relating to biobanking, where the nation state is no longer the central actor. As Andrew Barry has noted, biobanking can be considered to form a "technological zone" in which governance of activities and the establishment of standards operates beyond the traditional nation state borders.[33] Technological solutions, such as MIABIS, are seen as systems which address the problems that states are unable to adequately answer and solve. At the same time, however, the networks which operate across borders form new locations of governance and power through which information and data are channeled.

The BBMRI network also works toward establishing common services which its members can make use of. One example of this is the Common Service ELSI, which seeks to help discuss, establish, and foster ethical, legal, and social norms surrounding biobanking in the member states. Given the broad range of practices which have developed around biobanking during the decades in which it has been practiced, it has become important to develop common ethical and legal guidelines for operating and managing biobanks. Although many of the legal issues which biobanks face are also dictated by "hard law," such as the EU's data Directive, there are also a large amount of

---

[32] S. Tamminen, "Bio-Objectifying European Bodies: Standardisation of Biobanks in the Biobanking and Biomolecular Resources Research Infrastructure" (2015) 11 Life Sciences, Society and Policy 1.

[33] A. Barry, *Political Machines: Governing a Technological Society* (1st edn, Athlone Press 2001).

practices and procedures that can be guided by "soft law," or commonly agreed upon good practices within the biobanking community.

Apart from the joint efforts in biobank ontology summarized in MIABIS, BBMRI-ERIC has also organized working groups for the comparison and establishment of quality standards based on the CEN/TS. BBMRI-ERIC quality services provide the required tools, knowledge, and experience regarding quality management for biobanks and research on biomolecular resources.[34] These working groups have produced different self-assessment surveys or pre-examination processes for biobanks with the aim of evaluating quality of DNA, RNA, and proteins in various tissues, such as snap frozen and FFPE tissues, as well as for venous whole blood. They have also evaluated the prerequisites in metabolomics for urine, serum, and plasma, thus covering all possible preanalytical variables existing in the biobanked samples.

All these efforts, together with the development of IT tools such as sample/data locators and negotiators carried out by the BBMRI-ERIC common service IT, will pave the way to assist in and advise on the collection, distribution, and use of high-quality biobank samples. This will facilitate good practice for high-quality research, which will finally revert to the patient in precision diagnostics and personalized medicine. It is hoped that this will lead to better health and, therefore, quality of life.

5 DISCUSSION

In this chapter, we have tried to elucidate the importance of biobanks and biobank networks in biomedical research. Biobanks are central to the enactment and implementation of collection practices in which good quality samples and related information are collected from the population. This activity is central to the production of good research with results that are both replicable and valid. Maintaining and striving for good quality is also a valuable social goal since most of the large biobank collections, whether they be cohort studies, disease-based collections, or diagnostic collections in hospitals, have been collected and are maintained using public funding.

Biobank networks also serve an important function of giving rise to larger collections when these are combined. In the study of both complex diseases and rare conditions, it is necessary to have access to a sufficient number of samples to be able to derive meaningful conclusions. In many cases a single biobank is not enough. As such, biobank networks facilitate the process of

---

[34] BBMRI, "Quality Management" www.bbmri-eric.eu/BBMRI-ERIC/quality-management/ accessed June 4, 2018.

identifying which biobanks have the relevant samples that a researcher may need to conduct their study.

Ensuring both quality and quantity in biobanking should provide a more reliable avenue toward identifying and developing more accurate and clinically valid diagnostics and treatments for patients. As we have discussed, both quality and quantity are needed if we are to trust the biomarkers developed to identify disease causing genes, as well as gene–environment interactions. Without robust quality systems, as well as sufficient sample sizes, this may prove increasingly difficult and unreliable. Many of the purported benefits of personalized medicine are based on the assumption and goal that reliable and valid biomarkers are available for diagnostics and treatment. In order to facilitate research into personalized medicine it is necessary to provide the necessary research material, which can be used to study and validate possible biomarkers for personalized medicine.

In a broader sense, biobanks and biobank networks can be seen as technological implementations of political goals: in Europe, at least, the goal is to produce a more cohesive research market that will facilitate and accelerate the discovery of new biomedical innovations. Biobanks are also seen to serve as the basis of future innovations, as well as acting as drivers for the knowledge-based bioeconomy.[35] While facilitating these political aspirations, biobank networks also give rise to new governance structures in which new ethical and legal (soft law) norms are established and exercised. These norms have considerable implications in relation to our perceptions of the acceptability of new practices in relation to the collection, distribution, and use of biobanking samples. As such, biobanks and biobank networks play a crucial role in maintaining the social and technical norms which allow for tissue economies to emerge and function.

---

[35] A. Tupasela, "Data-Sharing Politics and the Logics of Competition in Biobanking" in Vincenzo Pavone and Joanna Goven (eds), *Bioeconomies: Life, Technology and Capital in the 21st Century* (Springer International Publishing 2017).

14. IP policies for large bioresources: the fiction, fantasy, and future of openness

Kathleen Liddell, Johnathon Liddicoat, and Matthew Jordan[1]

1 INTRODUCTION

Much research in synthetic biology (SB) and genomics (Gx) is reliant on the use of large scale collections of biological materials and data, often referred to as "biobanks" or "bioresources."[2] Following substantial investment in the form of time, money, and personnel, some of these bioresources have reached the point at which they can be regularly accessed by researchers and can realistically hope to facilitate innovation.[3] At the same time, their maturity brings with it several challenges, including how to promote access,[4] ensure

[1] The authors thank Linda Kahl, Biobricks Foundation, and Timo Minssen, University of Copenhagen, for stimulating discussions on the topic of this chapter.

[2] Johnathon E. Liddicoat and Kathleen Liddell, "Open innovation with large bioresources: goals, challenges and proposals" (2016) University of Cambridge Faculty of Law Research Paper No 6/2017. Available at SSRN: https://ssrn.com/abstract=2888871 accessed December 4, 2017.

[3] Eleni Zika et al, "A European survey on biobanks: trends and issues" (2011) 14(2) Public Health Genomics, 96; Saminda Pathmasiri, Mylène Deschênes, Yann Joly, Tara Mrejen, Francis Hemmings, and Bartha Maria Knoppers, "Intellectual property rights in publicly funded biobanks: much ado about nothing?" (2011) 29(4) Nature Biotechnology 319.

[4] Holger Langhof, Hannes Kahrass, Sören Sievers, and Daniel Strech, "Access policies in biobank research: what criteria do they include and how publicly available are they? A cross-sectional study" (2017) 25(3) European Journal of Human Genetics 293.

stewardship,[5] and address financial sustainability.[6] All these facets must be managed for ongoing utility. Access promotes the likelihood of significant scientific findings and avoids underutilization; stewardship earns trust from sample donors and funders; and sustainable sources of income are crucial if the bioresources are to serve as infrastructure (rather than projects) and assist with longitudinal studies.

A key part in managing these challenges—much less studied than issues of consent, return of incidental findings, and researchers' eligibility for access—is the stance taken by large bioresources on intellectual property (IP) and financial conditions of access. Acquisition, ownership, and sharing of IP in life sciences is ethically charged, and financial conditions of access are controversial where they preclude or discourage external researchers from using the bioresource. There is limited guidance available for developing such policies in the fields of Gx and SB.[7] Moreover—and this goes to the heart of this chapter—discussion of IP and business models for bioresources is a topic that falls into the shadow of hyperbole about openness.

This chapter aims to bring the issue of IP policies for large bioresources out of the long shadows of rhetoric about openness.[8] It will highlight two fictions: first, that the idea of openness is clearly defined; second, that organizations are committed to openness. It will also highlight the fantasy that harmonization of bioresources' access policies is feasible and desirable. The chapter will conclude by outlining future research to improve openness and IP policies for large Gx and SB bioresources.

---

5 Roger Brownsword, "Regulating biobanks: another triple bottom line" in Giovanni Pascuzzi, Umberto Izzo, and Matteo Macilotti (eds), *Comparative Issues in the Governance of Research Biobanks* (Springer 2013); Stephen J. O'Brien, "Stewardship of human biospecimens, DNA, genotype, and clinical data in the GWAS era" (2009) 10 Annual Review of Genomics and Human Genetics 193.

6 Don Chalmers et al, "Has the biobank bubble burst? Withstanding the challenges for sustainable biobanking in the digital era" (2016) 17(1) BMC Medical Ethics 39; Sally Gee, Rob Oliver, Julie Corfield, Luke Georghiou, and Martin Yuille, "Biobank finances: a socio-economic analysis and review" (2015) 13(6) Biopreservation and Biobanking 435.

7 Dianne Nicol and Richard Gold, "Standards for biobank access and intellectual property" in Matthew Rimmer and Alison McLennan (eds), *Intellectual Property and Emerging Technologies: The New Biology* (Edward Elgar Publishing 2012); Timo Minssen, Berthold Rutz, and Esther van Zimmeren, "Synthetic biology and intellectual property rights: six recommendations" (2015) 10(2) Biotechnology Journal 236.

8 In this article, the concept of "policies" refers to self-regulatory voluntary policies observed by bioresources, not legislative policies.

2 AN UNFORTUNATE FICTION

The concept of "openness" in bioresources is widely applauded, indeed reified, by scientists, governments, international institutions, and academic authors.[9] However, it is fictional in the sense that it is ill-defined and vague. One problem is that, within the existing literature, a variety of phrases endorsing openness are used interchangeably. These include "open science," "open innovation," "open source," "open access," "open biology," "open biotechnology," "open bioinformation," and so on. While these terms all refer in a broad manner to the sharing of information, they are not interchangeable. However, there is a tendency to use them as if they were. Take, for instance, "open biotechnology," which sometimes refers to a journal or database that can be accessed at no cost via the internet (but subject to copyright protection), and at other times refers to a biotech foundry where seeds can be sourced free of IP rights but subject to a fee.[10]

One task, to sharpen our understanding of openness in Gx and SB, is to clarify the various meanings. Pinpointing definitions is not straightforward because the terms are not used carefully or consistently by authors. Nevertheless, Table 14.1 shows a typology of openness, which is based on a wide-ranging literature review.[11] Given this chapter's interest in IP strategies for developing SB and Gx, the typology focuses on the ways in which openness-related vocabulary takes different attitudes toward different types of IP. It also has regard to the origins of the different vocabulary and typical business models associated with it.

---

[9] Westminster Higher Education Forum Keynote Seminar, "The next steps for delivering open access—implementation, expansion and international trends (forthcoming); Robert Cook-Deegan, Rachel A. Ankeny, and Kathryn Maxson Jones, "Sharing data to build a medical information commons: from Bermuda to the Global Alliance" (2017) 18 Annual Review of Genomics and Human Genetics 389; Nadine Levin and Sabina Leonelli, "How does one 'open' science? Questions of value in biological research" (2017) 42(2) Science, Technology, & Human Values 280; OECD, 'Making Open Science a Reality' (2015) OECD Science, Technology and Industry Policy Papers 25; Edward S. Dove, Vural Özdemir, and Yann Joly, "Harnessing omics sciences, population databases, and open innovation models for theranostics-guided drug discovery and development" (2012) 73(7) Drug Development Research 439; Arti K. Rai, "Open and collaborative research: a new model for biomedicine" in Robert W. Hahn (ed), *Intellectual Property Rights in Frontier Industries* (AEI Press 2005); HUGO, "Summary of principles agreed at the first international strategy meeting on human genome sequencing" (February 25–28, 1996) www.casimir.org.uk/storyfiles/64.0.summary_of_bermuda_principles.pdf accessed April 25, 2018.

[10] Yann Joly, "Open biotechnology: licenses needed" (2010) 28(5) Nature Biotechnology 417.

[11] Developed from Liddicoat and Liddell, *supra at* note 1.

Table 14.1 *Typology of openness vocabulary*

	Origins	Key feature	Accepting of IP?
Open science	Eco/soc science Observing norms of science (however, the practice of open science precedes this description by economists and social scientists).	Making research publications and data available. The development of knowledge that will not be developed by private individuals.	Copyright for recovery of costs. Patents are acceptable, but only in so far as private money is required to translate discoveries into market products. Use of trade secrets over data and information until publication.
Open access (A)	Governments/ universities/funding bodies	Digital availability of research publications.	Similar to "open science" in that copyright is accepted. But it is principally used to organize the relationship between the publisher and author, rather than as a means for recouping costs from readers. Generally speaking, "open access" focuses on access to data and publications and has nothing to do with patents or access to physical scientific materials.
Open source	Software industry	Online software development, allowing users to edit, use, and reuse source code.	Copyright to organize rules of access and use (including downstream use).
Open access (B) "Free software"	Software industry	Same as for open source.	No. Cultural norms for achieving open access.
Open patenting	Biotech industry	Allowing users to use patented DNA and organisms.	Patents to organize rules of access and use (including downstream).
Open innovation	Corporate organization theory (however, the practice of open innovation precedes this moniker).	Corporations collaborating. Looking beyond internal innovation.	Yes—all forms. Transfers of knowledge organized and incentivized via IP, money, and other tools. However, companies practicing open innovation will also look to for knowledge that is not protected by IP.

As Table 14.1 shows, the term "open source" is typically linked with free access to software source codes, and sees copyright as a tool for setting and enforcing rules to achieve that access. In contrast, "open science" is a norm observed by economists and sociologists to describe the culture among scientists to publish their discoveries and insights, assigning the knowledge to the community as a public good to be shared by all.[12] However, the knowledge is often kept confidential until scientists choose to reveal it. In addition, copyright is seen as a tool for recovering costs of publication, and as the natural moral right of scientists as authors.[13] Typically a fee is paid to read the article.

In contrast to both "open source" and "open science," the terminology of "open innovation," as coined by Henry Chesbrough,[14] describes a corporate innovation process. Companies adopting "open innovation" collaborate with other organizations, rather than innovating in a classically internal, secret, and siloed way. This form of openness is accepting of all types of IP rights *and* fees for access.

The additional terminology of "open access" introduces further confusion, because it is a phrase that appears to have two meanings. One meaning is derivative of "open science" and refers to the recent shift toward digital copies of scientific research articles being made available to readers for free, with the costs of publication being shouldered by the author or their employer/funder.[15] The other meaning of "open access" is an extension of the "open source" concept, or more specifically its cousin concept "free software." The free soft-

---

[12] Paul David, "The economic logic of 'open science' and the balance between private property rights and the public domain in scientific data and information: a primer" in Julie M. Esanu and Paul F. Uhlir (eds), *The Role of Scientific and Technical Data and Information in the Public Domain: Proceedings of a Symposium* (National Academia Press 2003), 19; Partha Dasgupta and Paul A. David, "Toward a new economics of science" (1994) 23(5) Research Policy 487.

[13] See for example Universal Declaration of Human Rights, Art. 27(2): "Everyone has the right to the protection of the moral and material interests resulting from any scientific, literary or artistic production of which he is the author."

[14] Henry Chesbrough, "The open innovation paradigm" in *Open Innovation: The New Imperative for Creating and Profiting from Technology* (Harvard Business School Press 2006).

[15] See for example European Research Council, "Open Access" https://erc.europa.eu/funding-and-grants/managing-project/open-access accessed December 4, 2017; Jean-Claude Guédon, "The 'green' and 'gold' roads to open access: the case for mixing and matching" (2004) 30(4) Serials Review 315; Peter Suber, "Creating an intellectual commons through open access" in Charlotte Hess and Elinor Ostrom (eds) *Understanding Knowledge as a Commons: From Theory to Practice* (MIT Press 2007); Bo-Christer Björk and David Solomon, "Open access versus subscription journals: a comparison of scientific impact" (2012) 10(73) BMC Medicine https://bmcmedicine.biomedcentral.com/articles/10.1186/1741-7015-10-73 accessed December 4, 2017.

ware movement distanced itself from the use of copyright and relies instead on behavioral norms for imposing rules of access.[16]

The contrast in the software industry between "open source" (*gratis*, usable but encumbered) and "open access" (*gratis* and unencumbered) is echoed in the life sciences sector. For instance, some commentators in the life sciences sector refer to openness as meaning arrangements inspired by "open source" ideas, where IP rights such as copyright coordinate rules of access or, in the case of "open patenting,"[17] where patents operate as the coordinating framework. With this approach, there are multiple discrete owners, who waive fees but impose terms on users via the legal authority they have as IP owners.[18] In contrast, where the terminology of "open access" is derivative of, or inspired by, the free software movement, it is based on "free revealing," which uses community norms and privileges (such as membership of research consortia, access to public funding, invitations to conferences) to maintain adherence. With this approach, innovation and discoveries are jointly owned, or stewarded, by the community. An example of a bioresource inspired by this approach is the Structural Genomics Consortium (SGC), which is aiming to catalyze research in new areas of drug discovery by focusing on understudied areas of human biology. SGC does not allow affiliated scientists to seek patents. Just as there are proponents of free software in the software sector who disagree with those favoring open source but see a greater foe in proprietary control, there are commentators in the life science sector who propose "open access" and disagree with open source, or open patenting. For example, Gold and Nicol suggest that "open access" would seem a better fit for biobanks—

---

[16] Richard M. Stallman, Lawrence Lessig, and Joshua Gay, *Free Software, Free Society: Selected Essays of Richard M. Stallman* (Free Software Foundation 2006); Richard Stallman, "Why open source misses the point of free software" www.gnu.org/philosophy/open-source-misses-the-point.en.html accessed December 4, 2017.

[17] Mariateresa Maggiolino and Maria Lillà Montagnani, "Standardized terms and conditions for open patenting" (2013) 14 Minnesota Journal of Law, Science and Technology 785; Mariateresa Maggiolino and Maria Lillà Montagnani, "From open source software to open patenting—what's new in the realm of openness?" (2011) 42(7) International Review of Intellectual Property and Competition Law 804.

[18] See E. Richard Gold and Dianne Nicol, "Beyond open source: patents, biobanks and sharing" in Giovanni Pascuzzi, Umberto Izzo and Matteo Macilotti (eds), *Comparative Issues in the Governance of Research Biobanks* (Springer 2013), 198–203; Dianne Nicol and Richard Gold, "Standards for biobank access and intellectual property" in Matthew Rimmer and Alison McLennan (eds), *Intellectual Property and Emerging Technologies—The New Biology* (Edward Elgar Publishing 2012), 142–4; and Mariateresa Maggiolino and Maria Lillà Montagnani, "From open source software to open patenting—what's new in the realm of openness?" (2011) 42(7) International Review of Intellectual Property and Competition Law 804, 811–14.

they view it as more efficient and less costly[19]—whereas Gitter, who seems to prefer a contractually organized "open source" approach, sees many flaws in the "open access" approach.[20]

We do not expect that people's use of language can be restricted to a lexicon of standard terms, but narratives of openness would be much improved if speakers and authors explained more clearly what they meant when referring to "open science," "open innovation," "open source," "open biotechnology," "open bioresources," "open patenting," and so on. Likewise, their audiences should consider critically what they actually mean, and whether a familiar phrase is actually being used in a different sense. When terms are used without clarification, discussions and debates quickly enter the world of fiction, where meaning is capable of many different interpretations, and subtext abounds but is easily overlooked.

3 A PROBLEMATIC FICTION

The next fiction is the way in which bioresources, governments, international organizations, and academic authors rhetorically iterate a commitment to "open" bioresources. This fiction is not only confusing but potentially harmful to the development of biomedicine and bioresources insofar as it undermines trust in ordinary life science business models. Semantically, their praise of "openness" seems to promise unencumbered access, and creates a climate where failing to laud "openness" or follow practices that protect private technical information leads to suspicion. Since the commercial, for-profit sector routinely protects private technical information as confidential information

---

[19] E. Richard Gold and Dianne Nicol, "Beyond open source: patents, biobanks and sharing" in Giovanni Pascuzzi, Umberto Izzo, and Matteo Macilotti (eds), *Comparative Issues in the Governance of Research Biobanks* (Springer 2013), 202; and Dianne Nicol and Richard Gold, "Standards for biobank access and intellectual property" in Matthew Rimmer and Alison McLennan (eds), *Intellectual Property and Emerging Technologies—The New Biology* (Edward Elgar Publishing 2012), 144.

[20] Donna M. Gitter, "The challenges of achieving open source sharing of biobank data" in Giovanni Pascuzzi, Umberto Izzo, and Matteo Macilotti (eds), *Comparative Issues in the Governance of Research Biobanks* (Springer 2013), 168. Donna Gitter examines barriers to widespread implementation of open source principles in biomedical research, and considers "fair access"/"restricted access" biobanks as an alternative. She identifies a number of the barriers against extensive openness in biobanking. These include: (i) reluctance among researchers to share their data; (ii) the challenge of crafting appropriate publication and IP policies; (iii) difficulties respecting informed consent, privacy, and confidentiality of research participants when data is shared so widely; (iv) controversy surrounding the issues of commercialization and benefit-sharing; and (v) complexity establishing a suitable infrastructure.

or trade secrets, this contributes to a culture of public distrust.[21] However, the reality is that when one investigates more deeply, most (if not all) bioresources, including large, publicly-funded bioresources, often employ what are better known as "controlled eligibility policies," which are not open in the sense of free from fees, free from conditions, or free from proprietary encumbrances. For instance, it is commonplace among Gx and SB bioresources to require payment for access;[22] to specify who can request material (including whether international transfers are possible); to insist on a declaration of intended purpose from the user; and to impose terms and conditions about ownership of existing and future data and materials, and IP rights.

A review of high-profile Gx and SB bioresources shows that there are exceedingly few instances of entirely free and unencumbered bioresources. The EMBL-EBI bioresource in the Gx sector comes close,[23] but even it does not meet the full promise of "open," free (*gratis* and unencumbered) knowledge and data flow. It allows users *gratis* access of its database without imposing its own constraints, but it does not guarantee freedom to use the data. Limitations on the use of data can be stipulated by the data "owners," who are not named or identifiable. Since it is not at all clear that data can be "owned" under English law,[24] it similarly unclear who might be the owners. It may be a reference to the depositors on the basis that they may have retained rights to control the entries through copyright or contract. In the SB sector, the Biobrick bioresource also comes close to being a *gratis*, unencumbered "open" bioresource. However, the Biobrick Public Agreement asserts a right of attribution for contributors, and a right for Biobrick to enforce fixation of the Biobrick trade mark if a user commercializes or distributes a part.

The bioresource policies just mentioned are actually outliers and represent some of the most "open" of access policies. Typically, "openness" polices are far more restrictive, especially when a bioresource is larger, is closer to clinical

---

[21] IPSOS Mori Social Research Institute, "The one-way mirror: public attitudes to commercial access to health data" (*Wellcome Trust*, 2016) www.ipsos.com/sites/default/files/publication/5200-03/sri-wellcome-trust-commercial-access-to-health-data.pdf accessed December 4, 2017.

[22] See for example Susan M.C. Gibbons, "Regulating biobanks: a twelve-point typological tool" (2009) 17(3) Medical Law Review 313, 342; Holger Langhof, Hannes Kahrass, Sören Sievers, and Daniel Strech, "Access policies in biobank research: what criteria do they include and how publicly available are they? A cross-sectional study" (2017) 25(3) European Journal of Human Genetics 293, 293.

[23] European Bioinformatics Institute, "Terms of use" www.ebi.ac.uk/about/terms-of-use accessed December 4, 2017.

[24] Tanya Aplin, Lionel Bently, Phillip Johnson, and Simon Malynicz, *Gurry on Breach of Confidence: The Protection of Confidential Information* (Oxford University Press 2012), 316; see also Art. 39 TRIPS.

applications, incorporates identifiable data, and/or is in the process of trying to achieve financial independence from research grants. More mixed policies might be better referred to as "closed" or "restricted" or "fair access," rather than open.[25]

For instance, access to UK Biobank (a collection of samples, health information, and genetic information from 500,000 people) is available only to "authorized" research projects,[26] and at a price.[27] Moreover, encumbrances are imposed on the users' results. For instance, contractually UK Biobank asserts what we call a right of enrichment, meaning that researchers are required to give back a copy of any new data they develop through the use of UK Biobank within 6 months of publication or 12 months of completing their project.[28] UK Biobank also asserts through a contract with users that it has a right to use the users' new datasets and to sublicense that data without payment.[29] UK Biobank also asserts a public interest march-in right, whereby it can insist that the IP rights of any kind developed through use of UK Biobank are licensed to it for free (and can be sublicensed by it) if the IP rights are being used in ways which UK Biobank deems "unreasonably restrictive."[30]

Another example of a highly mixed policy of openness and "closedness" comes from the UK's 100,000 Genome Project (GP), which hopes to include 100,000 fully sequenced human genomes.[31] Access to this database is not free for scientific researchers employed by commercial entities. They are required

---

25 UK DNA Banking Network (UDBN) explicitly rejected the unrestricted "open access" model, and instead developed a "fair access" regime which derives inspiration from the 2003 UNESCO International Declaration on Human Genetic Data. It distributes data via the project website to third party researchers, who apply for online registration. After verifying credentials, UDBN grants access to a restricted area of the website, where third party researchers can communicate online with the data collectors in order to negotiate collaboration. If negotiations are successful, UDBN then permits the data collector to grant the third party access to fuller data. This approach accepts a collector's right to "exclusive access to his/her collection for the purposes of the investigational goals stated in the initial collection proposal, recognising a 'first mover' advantage is likely to motivate scientific discovery."

26 UK Biobank, "Access procedures: application and review procedures for access to the UK Biobank resource" (2011) www.ukbiobank.ac.uk/wp-content/uploads/2011/11/Access_Procedures_Nov_2011.pdf accessed December 4, 2017, cl B1.5.

27 *Ibid* cl CA.1.

28 *Ibid* cl C11.4.

29 *Ibid* cl B8.4.

30 *Ibid* cl B8.5-B8.6.

31 As of November 6, 2017, the 100,000 Genomes Project had sequenced 39,540 genomes.

to pay a "fair and reasonable" "fee for service."[32] A variety of charging models is envisaged, including upfront payments and downstream royalties. While access to the database is *fee*-free for academic researchers, they must first join GeCIP (the Genomics England Clinical Interpretation Partnership), and membership has a potentially high price tag in the form of "reachthrough" rights for Genomics England (the private company established and wholly owned by the Department of Health that runs the bioresource on behalf of the UK state). The terms of the "reachthrough" rights are that a member who develops valuable results using only 100,000 GP project data must assign *all interests* in the results, including IP rights, to Genomics England.[33] The academic is granted back a *gratis* right to use the GeCIP Outputs, but *only* for noncommercial, academic research. Potentially, then, quite valuable results must be given to Genomics England by academic researchers in exchange for accessing the 100,000 GP. In different situations where the GeCIP member develops IP with substantial inputs from a source additional to the 100,000 GP, Genomics England's policies state that it will approach the situation on a case-by-case basis.[34] Genomics England has not yet publicly stated whether or not reachthrough rights apply to the results obtained by a commercial sector researcher, although commercial companies have been involved in some preliminary and precompetitive research work.

As these examples show, the reality is that "openness" is a spectrum ranging from "free (as in *gratis* and unencumbered) use of knowledge by anyone for any purpose" through to exclusive internal use by "one entity for its own purposes."[35] Notably, as shown by the examples just given, almost all access policies fall in the middle of the spectrum and are as closed or conditioned as they are "open."[36] There are a few reasons for access policies to fall

---

[32] Genomics England, "Intellectual property policy" (April 2017) cl 4.3, 16 www.genomicsengland.co.uk/about-gecip/for-gecip-members/documents/ accessed April 25, 2018.

[33] GeCIP, "Participation Agreement" (May 2017) cl 3.5, 7 www.genomicsengland.co.uk/about-gecip/for-gecip-members/documents/ accessed April 25, 2018.

[34] Genomics England, "Intellectual Property Policy" (April 2017) Annex cl 2.2, 10 www.genomicsengland.co.uk/about-gecip/for-gecip-members/documents/ accessed April 25, 2018.

[35] E. Richard Gold and Dianne Nicol, "Beyond open source: patents, biobanks and sharing" in Giovanni Pascuzzi, Umberto Izzo, and Matteo Macilotti (eds), *Comparative Issues in the Governance of Research Biobanks* (Springer 2013), 203. Gibbons clusters the spectrum of access models into three: (i) unlimited, free "open" or "public" access; (ii) "controlled" access (approved third parties, on application, subject to conditions); (iii) closed or exclusive access.

[36] See Andrew M. St. Laurent, *Understanding Open Source and Free Software Licensing: Guide to Navigating Licensing Issues in Existing & New Software* (O'Reilly Media 2004); Janet Hope, *Biobazaar: The Open Source Revolution and Biotechnology*

somewhere between completely open and completely closed. One reason is that open access policies sometimes impose specific compliance measures in order to avoid violation of the conditions of openness.[37] For example, "open source" software often has strings attached. While it may not be encumbered by private property rights, but it may be encumbered by specific compliance requirements, such as proper attribution to the copyright holder, recordkeeping requirements or downstream management issues.

Additionally, property rights can exist without going through special formalities such as registration. So even where a bioresource is committed to an extensive version of openness, the bioresource may nevertheless be aware that its data or samples might be encumbered by third party property rights. For example, in many jurisdictions, copyright—a form of property—attaches to original literary expressions (a low hurdle to pass) without a registration requirement. In some jurisdictions, such as EU member states, database rights also arise without registration. And in some jurisdictions, it is possible that third parties have property rights in samples of tissue. For example, case law has left open the possibility that individuals may have property rights in tissue that comes from their body or a deceased family member.[38] Furthermore, in English and US law it is well established that a scientist or clinician who puts skill and labor into a tissue sample (for example by preserving or developing it) can claim property rights over it.[39] There is also a robust legal discussion underway in many countries debating whether or not commercially valuable secret information (such as trade secrets or confidential information) can or should be protected as property.[40] This means that in a variety of situations, property rights regularly arise (and could potentially arise in a variety of

(Harvard University Press 2009); Mariateresa Maggiolino and Maria Lillà Montagnani, "From open source software to open patenting—what's new in the realm of openness?" (2011) 42(7) International Review of Intellectual Property and Competition Law 806, 806–7.

[37] Paul Liu, "Open source pitfalls at the biotechnology–high technology crossover" (2018) Thomas Jefferson School of Law Research Paper No 3189002. Available at SSRN: https://ssrn.com/abstract=3189002 accessed June 11, 2018.

[38] See for example Shawn Harmon and Graeme Laurie, "Yearworth v North Bristol NHS Trust: property, principles, precedents and paradigms" (2010) 69(3) Cambridge Law Journal 476, 478; and Kathleen Liddell and Jeffrey M. Skopek, "Informed consent for research using biospecimens, genetic information and other personal data" in *eLS* (John Wiley & Sons 2018), 7.

[39] For example Human Tissue Act 2004, s 32(9); Shawn Harmon and Graeme Laurie, "Yearworth v North Bristol NHS Trust: property, principles, precedents and paradigms" (2010) 69(3) Cambridge Law Journal 476, 481.

[40] OECD, "Enquiries into intellectual property's economic impact" (2015) 134, www.oecd.org/officialdocuments/publicdisplaydocumentpdf/?cote=DSTI/ICCP(2014)17/CHAP1/FINAL&docLanguage=En accessed April 25, 2018.

further situations, according to legal scholars) automatically through legal principles, rather than conscious acts of declaring and notice. To extinguish the ongoing existence of such property, in a way that users of a database can feel confident with, there must instead be a clear and traceable declaration. Unless extinguished in this way, they may continue or be perceived to continue, and it may be difficult to know the scope of the right, or to identify its owner.

Another reason why most bioresources end up with a mixed policy of openness and closedness is that fees, conditions on access, and march-in and reachthrough rights *can* be appropriate. It is incorrect to assume that everything open brings benefits, while closed approaches inevitably have negative effects.[41] The rapid decline in the fortunes of the newspaper and music industries provides a powerful reminder of the issues that can follow too much openness.[42] The music industry has largely recovered by implementing forms of closedness. Meanwhile, the newspaper industry is also trying to create new strategic digital positions to halt the hemorrhaging of its revenues.[43] Like these industries, biobanking is also trying to balance optimal readership and research use, continued pathways of innovation, and private sector involvement, while maintaining public trust and sufficient levels of funding.[44] A degree of closedness is even more likely with bioresources that include phenotypic data, full genome sequencing, and links to health records in order to appropriately manage privacy and data protection.[45]

---

41 Boston Consulting Group, "The new rules of openness" (2011) LibertyGlobal Policy Series, 3, 7. See also Morris W. Foster and Richard R Sharp, "Share and share alike: deciding how to distribute the scientific and social benefits of genomic data" (2007) 8 Nature Reviews Genetics 633, 635 cited by Gitter, "The challenges of achieving open source sharing of biobank data" in Giovanni Pascuzzi, Umberto Izzo, and Matteo Macilotti (eds), *Comparative Issues in the Governance of Research Biobanks* (Springer 2013) at 184, who says that promulgation of open access to biobanks "may not result in greater and more rapid scientific benefits" but may instead "result in duplication of effort, cause problems in the peer review system and create incentives for generating more publicly inaccessible databases, while reducing the number of biotech spinoffs from funded studies."

42 Boston Consulting Group, "The new rules of openness" (2011) LibertyGlobal Policy Series, 9.

43 *Ibid.*

44 Dianne Nicol and Richard Gold, "Standards for biobank access and intellectual property" in Matthew Rimmer and Alison McLennan, *Intellectual Property and Emerging Technologies—The New Biology* (Edward Elgar Publishing 2012), 136–8, 157.

45 "Restricted access databases" (databases that require authentication so that only bona fide researchers can obtain access) may be better suited to preserve privacy and confidentiality with respect to genomic databases. Restricted access also permits the database to provide some phenotypic information linked to genotypic data, enhancing the scientific value of the data: Timothy Caulfield et al, "Research ethics recommen-

Ideally, people involved with the development and critique of bioresources will take care to avoid the false presumption that openness is always better. This is an important foundation for a constructive discussion about appropriate conditions for the use and sharing of bioresources. This does not mean that the authors of this chapter oppose theses or campaigns that seek to shift the balance of openness and closedness toward greater openness. Not at all; we support several such initiatives. For instance, we support initiatives that make it possible for members of the public to read and refer to publicly-funded academic research papers without paying at the point of access. Our point is rather that it is important to drop the fiction that openness is absolute and unproblematic, and that it is always preferable. It is important to consider carefully whether a mix of terms and conditions is important, and the methods that will be used to meet the costs of sustaining a bioresource.

4 A FANTASY

There is a fantasy—in the sense of a desire for impossible or unlikely things—among bioresources, academic authors, and international organizations that bioresource access policies can be, and should be, consistent and harmonized.[46] Several large research projects have been given research funding to work on a harmonized approach.[47] Many authors critiquing biobanks move swiftly, sometimes imperceptibly, from the descriptive claim "there is much variability among biobank policies" to the normative claim "we need more harmonization."[48]

Rather differently, we would say that harmonization and consistency is a practice that must be carefully considered. In many situations, consistency

---

dations for whole-genome research: consensus statement" (2008) 6(3): e73 PLoS Biology, 434.

[46] For example, Healthcare Industry BW, "Medical ethicists: biobanks need harmonization rather than new laws" (June 2012) www.gesundheitsindustrie-bw.de/en/article/news/medical-ethicists-biobanks-need-harmonization-rather-than-new-laws/ accessed December 4, 2017.

[47] For example PHOEBE—Promoting Harmonisation of Epidemiological Biobanks in Europe (March 2015) www.fhi.no/en/projects/fp6---phoebe---promoting-harmonisat/ accessed December 4, 2017; and Public Population Project in Genomics and Society (P³G) www.p3g.org/about-p3g accessed December 4, 2017; see also Deborah Mascalzoni et al, "International charter of principles for sharing bio-specimens and data" (2015) 23(6) European Journal of Human Genetics 721.

[48] For example Holger Langhof, Hannes Kahrass, Sören Sievers, and Daniel Strech, "Access policies in biobank research: what criteria do they include and how publicly available are they? A cross-sectional study" (2017) 25(3) European Journal of Human Genetics 293; Timothy Caulfield et al, "A review of the key issues associated with the commercialization of biobanks" (2014) 1(1) Journal of Law and the Biosciences 94.

makes sense and contributes to operational efficiency, especially by reducing transaction costs. However, consistency is a not a virtue when one is faced with differences, adversity, or change. Maintaining the same beliefs, behavior, and policies—without considering whether and to what extent they are appropriate for the current situation—can cause problems.

In our view, it can be important to resist the siren call to harmonization.[49] As bioethicists such as Gibbons, and Chadwick and Strange point out,[50] the variability in bioresources policies is not only a byproduct of the heterogeneity of their design and purpose, but also a result of "legal and ethical doctrines ... remaining unsettled [and] ambiguous."[51] It has also been pointed out that if biobanks' access policies are uniform, this is a prima facie indicator that authors of the policies are failing to give due weight to plural views, empirical uncertainties, and legal variation.[52]

The speed of change among bioresources also hampers attempts to harmonize their access policies. An example of the rapidity of change can be seen in the changes between the holdings of UK Biobank and the younger 100,000 GP. The 100,000 GP prides itself on greater attention to DNA sequence accuracy, the availability of full genome sequences which allow genome-wide association studies, the involvement of patient participants who have diagnosed illnesses, and linked, curated NHS data.[53] These features increase the cost of running the bioresource and introduce somewhat different privacy issues.

Ironically, with all the attention given to the holy grail of harmonization, there are now multiple 'templates' for a harmonized access agreement for biobanks which are themselves not harmonized. For example, the International Charter of Principles for Sharing Bio-specimens and Data emphasizes the need

---

[49] Margoni cites a large body of literature on the pros and cons of standard form contracts: see footnote 20 in Thomas Margoni, "The roles of material transfer agreements in genetics databses and bio-banks" in Giovanni Pascuzzi, Umberto Izzo, and Matteo Macilotti (eds), *Comparative Issues in the Governance of Research Biobanks: Property, Privacy, Intellectual Property, and the Role of Technology* (Springer 2013), 241.

[50] Susan M.C. Gibbons, "Regulating biobanks: a twelve-point typological tool" (2009) 17(3) Medical Law Review 313. Ruth Chadwick and Heather Strange, "Biobanking across borders: the challenges of harmonisation" in Deborah Mascalzoni (ed), *Ethics, Law and Governance of Biobanking: National, European and International Approaches* (Springer 2015), 133–8.

[51] *Ibid* at 341.

[52] Emmanuelle Rial-Sebbag and Anne Cambon-Thomsen, "The emergence of biobanks in the legal landscape: towards a new model of governance" (2012) 39(1) Journal of Law and Society 113, 119, 127–8.

[53] Genomics England, "The 100,000 Genomes Project Protocol" (January 2017) 6–9 www.genomicsengland.co.uk/100000-genomes-project-protocol/ accessed April 25, 2018.

to protect the IP of those who *create* biobanks.[54] This is on the grounds that they expend a lot of time and resources in creating the bioresource. In contrast, the template developed by the National Cancer Research Institute (citing the Confederation of Cancer Biobanks) states that donors and creators of biobanks should *waive* their IP rights, whereas IP needs to be carefully protected by the *users* of the bioresource,[55] as it is they who likely need exclusivity to develop marketable applications. In contrast to both, the P3G Generic Template falls into a trap of finding a "harmonized" principle which might have a reasonable claim to consensus by erasing most of the content from the principle. Its proposal is more or less that bioresource access policies should implement a transparent and balanced IP policy.[56]

One further issue is that harmonization is particularly difficult and fragile in the field of IP law, which, despite various international agreements, is still enforced at a national level, and is subject to an immense variety of licensing practices. Furthermore, in the field of IP, harmonization unravels much more quickly than it is generated, because it is a lightning rod for international politics and power plays for economic advantage.[57]

Our recommendation is that rather than awkwardly forcing harmonization upon bioresources, we should instead seek a shift in user expectations, acknowledging the fact that consistency is not always a virtue (depending on circumstances) and that IP and access policies are not, and should not be, necessarily designed to apply uniformly.

Reconceptualizing away from "open access" and toward "reasonable and tailored access" both addresses the problem of inconsistent terminology and makes clear that the access that is offered is thought to be reasonable in the particular circumstances, and may differ from bioresource to bioresource and from applicant to applicant. It recognizes that the interests and goals of all

---

[54] Deborah Mascalzoni et al, "International Charter of principles for sharing bio-specimens and data" (2015) 23(6) European Journal of Human Genetics 721, 722–4, 726.

[55] National Cancer Research Institute, "Samples and data for research: template for access policy development" (June 2009) 23 www.ncri.org.uk/wp-content/uploads/2013/09/Initiatives-Biobanking-2-Access-template.pdf accessed December 4, 2017.

[56] Ariane Mallette, Anne Marie Tassé, and Bartha Maria Knoppers, "P3G model framework for biobank governance" (2013) 10 www.p3g.org/system/files/biobank_toolkit_documents/P3G%20Model%20Framework%20for%20Biobank%20Governance%20FINAL%20%281%29_0.pdf accessed December 4, 2014.

[57] Henning Grosse Ruse-Khan, *The Protection of Intellectual Property in International Law* (Oxford Univeirsty Press 2016), 6, 9, 22; see also Martti Koskenniemi, "Fragmentation of international law: difficulties arising from the diversification and expansion of international law: report of the Study Group of the International Law Commission" (2006) UN Doc A/CN.4/L.682, 244–5.

stakeholders will vary, and so promises a more "tailored" approach to IP and access policies to take this into account.

In our view, it is very unlikely that specific rules can be agreed for a generally applicable IP policy for large bioresources, considering the substantial heterogeneity among the "mixed" systems of open access. It is possible that a degree of standardized tailoring—less than bespoke and more than one size fits all—could be attempted, but only once appropriate subgroup clusters have been identified. Continuing the tailoring metaphor, a series of standard sizes might be offered (similar to the way that clothing is clustered into sizes such as S, M, L, XL) to particular clusters of biobanks. However, identifying appropriate clusters—where the similarities hold relevant and sufficient force for a singular access policy, and the differences do not—will not be easy. The size of sample holdings may be one issue, but there will be many other issues influencing whether a particular access policy "fits" a biobank. Examples include available sources of funding, the purpose of the bioresource, the origins of samples, the terms of donors' consent, the maturity of the biobank (for example, is it in a phase of creation, collection or access?),[58] whether the contents are finite (as in the case of tissue samples) or infinite (immortal cell lines, data), and the annual cost of maintenance (which in turn may be related to whether there are tissue holdings, clinical grade genetic sequencing, dynamic record linkage, and so on).

5 THE FUTURE

To recap, our analysis of the literature identifies an unhelpful degree of hyperbole about the idea that biobanks should be open or else science suffers. It also identifies a rhetorical attitude toward harmonization, with a tendency to presume, without supporting evidence, that the simplicity of harmonization outweighs the precision of tailoring. Drawing attention to these biases allows IP policies to be discussed in a more constructive light, rather than face immediate distrust and suspicion for being at odds with "openness."

From this position, the next step is to investigate and debate the sorts of IP policies that would be useful for large Gx and SB biobanks to adopt. Like others before us, we think that a rich vein of sociolegal information can be found by surveying existing bioresource access policies to ascertain the variety of different policies currently offered.

---

[58] Saminda Pathmasiri, Mylène Deschênes, Yann Joly, Tara Mrejen, Francis Hemmings, and Bartha Maria Knoppers, "Intellectual property rights in publicly funded biobanks: much ado about nothing?" (2011) 29(4) Nature Biotechnology 319.

Table 14.2 *Principal empirical surveys of human bioresources' access policies*

Author	Year of Publication	Size of survey (descending)	Scope
Zika et al[1]	2010	126 human biobanks	Technical challenges, regulatory requirements, funding, finance, public engagement. Not IP policies
Joly et al[2]	2015	114 past/potential donors	Inter alia, donors' opinions on open science
Langhof et al[3]	2016	74 access policies for human biobanks	62 criteria for controlled access, but not IP policies
Capacosa et al[4]	2016	46 human biobanks	Regulatory requirements, access criteria, funding. Not IP policies
Capron et al[5]	2009	87 experts in human biobanking	Ownership, regulation, collective consent, benefit sharing. Not IP policies
Joly et al[6]	2011	16 documents/ incidents from human biobanks	Sanctions for noncompliance with access policies
Gibson et al[7]	2017	21 Canadian human biobanks	Access policies
Shabani and Borry[8]	2016	20 interviews with Data Access Committees	Key elements in controlled access
Lemrow et al[9]	2007	14 access policies	Nine key elements in controlled access (brief on IP)
*Knoppers et al[10]	2013	10 large publicly funded human bioresources	Access principles and norms including, inter alia, enrichment, ownership, IP, cost recovery
Gee et al[11]		8 human biobanks	Biobanking revenue models
*Nicol and Gold[12]	2012	3 human biobanks	IP-related standards

Notes:
[1] Eleni Zika et al, "A European survey on biobanks: trends and issues" (2011) 14(2) Public Health Genomics, 96.
[2] Yann Joly et al, "Fair shares and sharing fairly: a survey of public views on open science, informed consent and participatory research in biobanking" (2015) 10 PLoS ONE, e0129893.
[3] Holger Langhof, Hannes Kahrass, Sören Sievers, and Daniel Strech, "Access policies in biobank research: what criteria do they include and how publicly available are they? A cross-sectional study" (2017) 25(3) European Journal of Human Genetics 293.
[4] Marco Capocasa et al, 'Samples and data accessibility in research biobanks: an explorative survey.' (2016) 4 PeerJ e1613.

[5] Alexander Morgan Capron et al, "Ethical norms and the international governance of genetic databases and biobanks: findings from an international study" (2009) 19(2) Kennedy Institute of Ethics Journal 101.
[6] Yann Joly et al, "Genomic databases access agreements: legal validity and possible sanctions" (2011) 130(3) Human Genetics 441.
[7] Shannon G. Gibson et al, "Transparency of biobank access in Canada: an assessment of industry access and the availability of information on access policies and resulting research" (2017) 12(5) Journal of Empirical Research on Human Research Ethics 310.
[8] Mahsa Shabani and Pascal Borry, "'You want the right amount of oversight': interviews with data access committee members and experts on genomic data access" (2016) 18(9) Genetics in Medicine 892.
[9] Shannon M. Lemrow et al, "Key elements of access policies for biorepositories associated with population science research" (2007) 16(8) Cancer Epidemiology Biomarkers & Prevention 1533.
[10] Bartha Maria Knoppers et al, "A P3G generic access agreement for population genomic studies" (2013) 31(5) Nature Biotechnology 384.
[11] Sally Gee et al, "Biobank finances: a socio-economic analysis and review" (2015) 13(6) Biopreservation and Biobanking 435.
[12] Dianne Nicol and Richard Gold, "Standards for biobank access and intellectual property" in Matthew Rimmer and Alison McLennan (eds) *Intellectual Property and Emerging Technologies: The New Biology* (Edward Elgar Publishing 2012).

Table 14.2 summarizes the leading empirical surveys of human bioresources identified through literature reviews. Rather than organizing the results chronologically, the table is organized in descending order according to the number of biobanks studied in each publication. The largest was the study by Eleni Zika and others, which looked at 126 human biobanks. Two points emerge. First, it becomes clear that very few of the previous studies looked at the IP policies of bioresources. Instead, research has focused on: (i) access "eligibility" criteria (for example, who is allowed access, and what forms, information, and processes scientists need to complete); (ii) regulatory requirements (what rules must a bioresource follow with respect to donor consent, return of findings); and (iii) fees. Second, in the small number of articles which have analyzed IP policies—Knoppers et al and Nicol and Gold—the number of human bioresources investigated has been very small. Knoppers and colleagues studied ten large biobanks, all funded by public money. Nicol and Gold reviewed three large, public biobanks. It is worth noting that Perry investigated 31 biobanks but none of these focused on *human* biospecimens and data, but rather plants and animal bioresources.[59]

Looking to the future, it is important to address the missing information about IP policies for human bioresources. We recommend focusing on subsectors of the bioresources industry that are generating commercially valuable IP,

[59] Mark Perry, "Accessing accessions: biobanks and benefit-sharing" in Giovanni Pascuzzi, Umberto Izzo, and Matteo Macilotti (eds), *Comparative Issues in the Governance of Research Biobanks: Property, Privacy, Intellectual Property, and the Role of Technology* (Springer 2013).

or are approaching that point.[60] In these subsectors, one is more likely to find bioresources that are keen to understand possible options better because they regard IP as potentially valuable, and bioresources which have taken some initiative in drafting and trialing IP policies. For instance, the 100,000 GP has been quite proactive in designing an IP policy, and is keen to reflect critically upon, and potentially amend, the choices it has made to date. The 100,000 GP is large, aimed at finding clinical or end-user applications (not merely researchworthy results), and utilized by for-profit and nonprofit organizations. One significant challenge in this research will be to draw out meaningful information between the characteristics of different bioresources and the IP policy variables they adopt (Table 14.3).

---

[60] In 2011, in a paper subtitled "Much Ado about Nothing?" (cited in note 2), Pathmasiri et al argued that IP policies were not of great importance, as at that time very few human biobanks had been fully established to the point where scientists were generating new knowledge and inventions (they were still in the stage of set up). Further, the IP attaching to the data and samples was of limited value. In 2005, John Walsh and colleagues found that IP-related issues were (at least as a matter of self-report) relatively unimportant for scientists' decisionmaking about which research projects to pursue compared with concerns about scientific importance, feasibility, funding and material transfer agreements: John P. Walsh, Charlene Cho, and Wesley M. Cohen, "Patents, material transfers and access to research inputs in biomedical research. Final Report to the National Academy of Sciences' Committee Intellectual Property Rights in Genomic and Protein-Related Inventions" (September 2005) http://citeseerx.ist.psu.edu/viewdoc/download?doi=10.1.1.5 31.1401&rep=rep1&type=pdf accessed December 4, 2017. Since the publication of those articles a number of years have passed and the biobanking landscape has changed dramatically. It is time to review the situation. Notably, the financial environment is tougher, since there is now much less public funding available to maintain bioresources: Don Chalmers et al, "Has the biobank bubble burst? Withstanding the challenges for sustainable biobanking in the digital era" (2016) 17(1) BMC Medical Ethics 39. As a result self-funding through revenue raising has come to the forefront of discussions. Whether or not, and in what way, IP rights could play a role in this is a question worthy of investigation. Another significant change since the Pathmasiri and Walsh articles is the seismic changes in US patentability law following Association for Molecular Pathology v. Myriad Genetics, Inc., 133 S. Ct. 2107 (2013), Mayo Collaborative Services v. Prometheus Laboratories, Inc., 132 S. Ct. 1289 (2012), and Alice Corporation v. CLS Bank International, 132 S. Ct. 2347 (2014). *Mayo* in particular increased the divergence between US and European IP law.

Table 14.3 Focuses for empirical survey of bioresources IP policies

Characteristics of bioresource	IP policy variables
Biology	Definition of IP
Size	Ownership asserted or reserved?
Scope	Nonchallenge clause[1]
Country location	Reachthrough clause
Commercial status	Attribution clause
Age/intended duration	Enrichment clause[2]
Content—DNA, data, tissue, health records?	March-in clause
Network membership	Controlled publication
Public funding?	Royalties
Cost recovery model	Viral clause cascading terms to onward licensing and assignment
	Sanctions[3]

Notes:
[1] Also known as "patent peace" or "retaliation" clause: the clause provides that the license will terminate if the licensee initiates litigation (including a cross-claim or counterclaim) alleging that a patent claim is infringed.
[2] Also known as "grant back."
[3] Yann Joly, Nik Zeps, and Bartha M. Knoppers, 'Genomic databases access agreements: legal validity and possible sanctions' (2011) 130(3) Human Genetics 441.

Notwithstanding US Supreme Court decisions such as *Myriad* and *Prometheus*, there is still a significant amount of activity in the area of gene-related patent applications and grants.[61] Relatedly, there is reason to think that the bioresources and their users will be interested in the prospect of patents and using them to translate scientific breakthroughs into products and therapies. There is also reason to think they will be concerned about future licensing, revenue, signaling, data aggregation, and exclusivity opportunities. These are precisely the sorts of points that a bioresource IP policy should seek to address. More specifically, bioresources ought to consider and address the issues listed in the *right-hand* column of Table 14.3. Furthermore, in order to address the question of whether bioresource IP policies might be usefully clustered, it is worth investigating whether the characteristics in the first column correlate with different sorts of clauses in bioresource IP policies in the second column.

An empirical survey of the type just described will not answer normative questions such as whether or not a biobank's IP policy should include downstream reachthrough rights—studying the world as it is will not tell us how it should be. Nevertheless, it has the potential to provide some useful and illumi-

[61] Mateo Aboy, Kathleen Liddell, Jonathon Liddicoat, and Cristina Crespo, "Myriad's impact on gene patents" (2016) 34(11) Nature Biotechnology 1119.

nating data about the variety of different IP approaches and the main drivers and trends, and it can also help identify case studies which can then shed light on how the policies are working in practice. In theory, it may even form an evidence base for refinement of future access policies.

Outside the biobank sector there are useful precedent studies. Notably, Lévesque and colleagues investigated 94 funding and policy guidance documents (in three languages) related to human embryonic stem cell research.[62] The documents covered basic issues related to open access and commercialization. For example, the authors investigated whether they embedded clear proprietary or open science requirements in their models for sharing eHSC material, data, and knowledge. The authors also investigated whether there was funding available to assist commercialization, or to assist with IP rights. Two other studies seeking to understand the diversity of openness and closedness in the life sciences non-biobank sector include Pressman and colleagues' study of variable licensing practices with gene-related patents,[63] and Margoni's study of the variable constraints imposed by MTAs at Canadian private and public research centers.[64]

6 CONCLUSION

Given the siloed and proprietary approach to scientific research early in the twentieth century, it is pleasing that a culture of openness has emerged and been adopted in SB and Gx; private and proprietary approaches can be too rigid, can arguably be unethical (when people are denied access to the scientific commons), and can involve excessive transaction costs. However, in the world of bioresources, the hyperbole surrounding the idea of openness confuses its application and utility. There needs to be more clarity about intended meanings when using the vocabulary of openness. There are many different forms of openness. We also need to soften the prevailing attitude that "open is good and closed is bad." Openness is rarely absolute; indeed, most bioresources adopt hybrid policies of controlled access rather than provide their scientific knowl-

62 Maroussia Lévesque, Jihyun Rosel Kim, Rosario Isasi, Bartha Maria Knoppers, Aurora Plomer, and Yann Joly, "Stem cell research funding policies and dynamic innovation: a survey of open access and commercialization requirements" (2014) 10(4) Stem Cell Reviews and Reports 455.

63 Lori Pressman et al, "The licensing of DNA patents by US academic institutions: an empirical survey" (2006) 24(1) Nature Biotechnology 31.

64 Thomas Margoni, "The roles of material transfer agreements in genetics databses and bio-banks" in Giovanni Pascuzzi, Umberto Izzo, and Matteo Macilotti (eds), *Comparative Issues in the Governance of Research Biobanks: Property, Privacy, Intellectual Property, and the Role of Technology* (Springer 2013), 241.

edge and ideas free from fees and unencumbered by third party legal rights or requirements, and there can be good reasons for a blended approach.

This chapter has also challenged the rhetorical view that bioresources should be working toward harmonized policies of openness. Legal and ethical harmonization in the sense of uniformity is only appropriate in "like" cases, and even then it may not be viable where legal systems differ. Rather than force awkward harmonization or vacuous principles upon a bioresource, it would be better to change bioresource users' expectations so that variability of openness and IP policies can be more easily accepted.

It is also important that more research be conducted to build a better evidence base for the future of openness and IP policies. A first step would be to identify the variety of approaches currently adopted and being trialed, for instance, the various definitions of IP utilized by biobanks, and instances of clauses stipulating nonchallenge, reachthrough, attribution, march-in, and enrichment requirements. Human biobanks seeking to find medical breakthroughs are obviously an interesting site to study, but it may be that biobanks dealing with animal, plants, or microorganisms are more "mature" and so have more experience of managing IP arrangements. With a better grasp of the approaches that could be taken, debates can move on to questions about how to decide between the options.

Index